MINING FOR SUSTAINABLE DEVELOPMENT IN TANZANIA

*Kassim Kulindwa, Oswald Mashindano,
Fanuel Shechambo & Hussein Sosovele*

DAR ES SALAAM UNIVERSITY PRESS LTD.

Published by:
Dar es Salaam University Press Ltd.
P.O. Box 35182
Dar es Salaam
TANZANIA

ISBN 9976 60 390 8

CONTENTS

ACRONYMS

AAPS	-	Africa-Asia Precious Stones
AIDS	-	Acquired Immunity Deficiency Syndrome
BOT	-	Bank of Tanzania
CT	-	Claim Title
EIA	-	Environmental Impact Assessment
ERB	-	Economic Research Bureau
ERP	-	Economic Recovery Programme
ESAP	-	Economic and Social Action Programme
FEMATA	-	Federation of Miners Associations in Tanzania
GDP	-	Gross Domestic Product
ICBT	-	Illegal Cross Border Trade
IMF	-	International Monetary Fund
IPC	-	Investment Promotion Centre
MC	-	Marginal Cost
MDL	-	Mineral Dealer Licence
MPs	-	Members of Parliament
MR	-	Marginal Revenue
NCPI	-	National Consumer Price Index
NEMC	-	National Environmental Management Council
NESP	-	National Economic Survival Programme
NGO	-	Non Governmental Organisation
NIPPA	-	National Investment (Promotion and Protection) Act
RPFB	-	Rolling Plan and Forward Budget
SADC	-	Southern Africa Development Community
SAP	-	Structural Adjustment Programme
SSM	-	Small-Scale Miners
STAMICO	-	State Mining Corporation
STDs	-	Sexually-Transmitted Diseases
TAMIDA	-	Tanzania Mineral Dealers Association
TCM	-	Tanzania Chamber of Mines
TET	-	Tanzania Economic Trends
TGI	-	Tanzania Gemstone Industry

TIC	-	Tanzania Investment Centre
TRA	-	Tanzania Revenue Authority
UDSM	-	University of Dar es Salaam
UNDP	-	United Nations Development Programme
URT	-	United Republic of Tanzania
US$	-	United States Dollar
USAID	-	United States Agency for International Development
WWF	-	World Wildlife Fund

List of Tables

List of Maps

List of Boxes

List of Figures

List of Table Appendeces

ACKNOWLEDGEMENTS

The study on Tourism for Sustainable Development in Tanzania was initiated by the Macroeconomic Programme Office of the World Wide Fund for Nature of the United States of America (WWF-US/MPO) in collaboration with the Economic Research Bureau (ERB) of the University of Dar es Salaam. This study is among several case studies, which have been conducted in four Southern African countries of the Republic of South Africa (RSA), United Republic of Tanzania (URT), Zambia and Zimbabwe under the, "Macroeconomic Reforms and Sustainable Development in Southern Africa Project."

This project was conducted and managed with the assistance of several bodies including the National Advisory Committee (NAC), Donor Forum (DF), Informal Stakeholder Forum (ISF), The Regional Seminars in Zanzibar, Lusaka-Zambia and Harare-Zimbabwe and the International Advisory Committee (IAC). Quite a number of individuals and institutions therefore have been involved in bringing this study to fruition.

First, we want to express our heartfelt gratitude to the WWF-US for making this study possible in Tanzania through co-ordination of resource mobilisation effort and management of the project in the four countries. Here we would particularly like to thank Dr. David Reed, the Director of MPO for providing us with the opportunity to work on this important and challenging topic. Thanks are also due to Ola Larsson, Programme Officer WWF-US for his able and superb facilitation.

At the national level, we would like to express our appreciation for the guidance and support given by the members of the NAC as follows: Mr. Eric Mugurusi, Director, Division of the Environment-The Vice President's Office, Mr. C. Chiragi, Ministry of Natural Resources and Tourism (MNRT), Mr Ramadhani Kijjah-Ministry of Finance, Mr. P.B. Rweyemamu- President's Office Planning Commission, Mr. Edwin Ngonyani-Ministry of Energy and Minerals, Mr. A.G.T. Nyenza-Ministry of Water, Mr. S.J. Makundi-Ministry of Land, Mr. I.A.Mchallo-National Environmental Management Council (NEMC), Dr. Paul Siegel, the Country Representative, the WWF Tanzania Country Office, Mr. P.Y. Nnyiti-Wildlife Conservation Society of Tanzania (WCST), Mr. Eric Munisi, Federation of Miners in Tanzania (FEMATA), Mrs. S. Massawe,Tanzania Tourist Board (TTB), Ms. Sango Kipozi-World Radio for Environmental Network (WREN), Mr. J. Mpinga- Journalists for the Environment in Tanzania (JET), Dr. Ammon Mbelle - Department of Economics, University of Dar es Salaam, Prof. S.L Chachage - Department of Sociology, University of Dar es Salaam.

Apart from the NAC, the Donor Forum shared with us its experience and provided appropriate guidance towards the successful accomplishment of this important task. The members include Mrs. Dorothy Hope Nsherenguzi (Royal Danish Embassy), Dr. Rolf Drescher, (Head, Division for Development Cooperation and Commercial Affairs, German Embassy), Mr William Michel (CIDA), R. Sabatucci (EU), Thomas Andersson (Sida), Olivier Burki (SDC) and Peter Dewees (World Bank).

At the regional level, we would like to thank our fellow partners in the project from RSA, Zambia and Zimbabwe and also key participants to the regional workshops conducted in Zanzibar, Harare and Lusaka. Here we particularly thank Dr. Jan Bojo (World Bank), Mr. Samuel Landon (CIDA), Mr. Michael Goldblatt (Development Bank of South Africa-DBSA), Mr. Glynn Davies (DBSA), Dr. Jorgen Levin (Swedish Embassy), Mr. Ali Hanein (Ministry of Tourism - Zanzibar), Mr. Muhammad Sulaiman (Ministry of Water, Construction, Lands and Environment), Mr. Rabson Dhlodhlo (ZERO - Zimbabwe), Dr. Ramos Mabugu (University of Zimbabwe), Dr. Guy Scott (MANO - Zambia), Ms Charlotte Harland (MANO - Zambia).

We would also like to convey our sincere appreciation to the International Advisory Committee members. These provided invaluable guidance through their participation during the advisory committee meetings in Brussels by providing us with challenging comments and advice, which have helped greatly in putting the work in the right perspective and context of the global situation. We particularly thank Mr. Arthur Runge-Metzger (European Commission, DGVIII/A/6, Office S/10), Ms. Nadine Speich (Environment, Forest Energy Division, Swiss Agency for Development and Cooperation-SDC), Ms. Maria Savvaides-Polyzou (European Commission, Head of Division DGVIII/A/6, Office 5/05), Mr. Francois Falloux (Principal Enviroment Specialist, World Bank), Mr. Eckehard Weiss (Project Officer/ Deutsche Gesellschaft fur Technische Zusammenarbeit (GTZ)).

We would further like to acknowledge with appreciation the contribution of the participants to a project presentation meeting at the European Commission in Brussels. These are Mr. Moses, T. Kibwana (Embassy of Tanzania, Brussels), Edward C. Sefuke (Embassy of Zambia, Brussels), Beatrice Mutetwa (Embassy of Zimbabwe, Brussels), Arthur Runge-Metzger (European Commission/DGVIII/A/6), Philip Lowe (European Commission/ GDVIII), Jacob Houtman (European Commission/DGVIII/A), Lad Biesebrock (European Commission/DGVIII/A), F. Bang-Hansen (European Commission /DGVIII/A/3), Maria Savvaides-Polyzou (European Commission/DGVIII/A/6), Amos Tincani (European Commission/DGVIII/ B3), Joao Ferreira (European Commission/DGVIII/E/3), Roger Moore (European Commission/DGVIII/E/3), Raffaela Fazio and Sascha Picher (European Network on Debt and Development), Bernadetto Nee (FOS/ ERO, Brussels), Tony Long (WWF/European Policy Office).

Furthermore, we would like to thank our fellow researchers from the three countries for their contribution in the discussions during the Regional Workshops conducted in Zanzibar, Harare and Lusaka, which greatly helped to improve the quality of this work.

In Tanzania, we are thankful to our employer, the University of Dar es Salaam through the Economic Research Bureau (ERB), and the Institute of Resource Assessment where we work, for allowing us to take time off to perform this important study and for their institutional support.

The fieldwork could never have been accomplished the way it did without the able participation of our research assistants. Here we would like to convey our sincere gratitude to Mathew Frank Mwamsamali, Athumani Faridi, and the late Mr. A.S. Kapele for their perseverance and hard work.

In the research areas of Ngorongoro, Serengeti, Bagamoyo, Arusha, Zanzibar, Kilimanjaro, government officials in the department of natural resources, economic planning, village leaders, ordinary villagers (respondents) made the field survey possible. Without them, data used for this analysis would have not been obtained. We thank them for their co-operation, understanding and patience during the survey.

We owe special thanks to Dr. A.V.Y. Mbelle for his contribution in reviewing the manuscript and providing us with valuable and constructive comments, which have further improved the quality of the study. Ms. Catherine Malwa of the Economic Research Bureau, was the project's administrative secretary who was responsible for procurement, word processing of the questionnaires and various versions of the reports. We thank her for her resilience. We also thank Mr. Hillary Mambo, the accountant for the Economic Research Bureau who was responsible for the project's financial management.

We also acknowledge the hardwork and professional conduct of the drivers who took the research teams around the study areas. Here we are grateful to Mr. Salum Kibode of ERB and Augustino Yona of IRA.

Finally, we thank Mr. John W. Mgonja for facilitating the preparation of project materials at various stages.

Despite all the cooperation, contributions and guidance accorded to us, the responsibility for any errors and deficiencies remain with the authors alone.

Dr. Kassim Kulindwa
Dr. Oswald Mashindano
Dr. Fanuel Shechambo
Dr. Hussein Sosovele

Dar es Salaam
December 2002

A SYNTHESIS OF EMPIRICAL FINDINGS

The main objective of this study is to analyse how macro-economic reforms in Tanzania and the policy instruments applied to achieve the objectives are faring in the mineral sector i.e. the response of the key players in the mineral sector. This is followed by an investigation into the impact on the economy, society and the environment. The intention is to establish whether or not the expanding activities can ensure a sustainable path for Tanzania. If not, then the results could be used to inform a policy dialogue that will lead to a review of policies and practices that would steer the economy along a more sustainable development path.

A case study approach is used, whereby three study sites were purposely selected. These are Mererani in Arusha, famous for tanzanite and other gemstones. The second study site is Geita in Mwanza Region, famous for gold mining by small artisanal miners as well as large-scale miners, mostly foreign. The third area is Umba Valley in Tanga Region, famous for its gemstones such as sapphires, green and red garnets, tourmalines, etc. Both secondary and primary data were collected, the former from government reports and other publications. This formed the literature review part of the study. Key informants were interviewed in regional mining offices, mining associations, mineral dealers and other interested persons, individually and corporately. A sample survey was carried out in the three mining areas, in which 201 persons were interviewed.

The starting point for the study is the macroeconomic reforms undertaken since 1986. Various economic instruments have been deployed to try and achieve the following objectives: achieving and sustaining macroeconomic stability, liberalisation of trade and foreign exchange regime; combating poverty and improving the welfare of the people. Others include creating a strong private sector and reducing government involvement in directly productive activities, improving efficiency in the use of public resources and achievement of a sustainable development path. The outcomes of these reforms have been diverse.

Economic Issues Arising include the following:

The outcome of these reforms has been a remarkable expansion of the mineral sector in terms of increasing numbers of mining companies, participation of artisanal miners, increased number of claim holders, growth in mineral settlements and towns as well as auxiliary activities (service provision). This expansion is having an impact on the economy, the environment as well as the social sphere, which we ought to follow up to assess the extent to which they lead the country towards sustainable development.

The benefits can be gauged in terms of economic growth, creation of employment, generation of tax revenues, increase in exports and economic growth resulting from a larger contribution of mining to GDP over time. According to the Federation of Miners Associations in Tanzania (FEMATA), more than 1,000,000 individuals are involved in mining activities.

However, the type of employment is informal, with all its problems of tax collection, improper contractual arrangements and thus a source of conflict between the "employees" and the "employers."

The tax regime for the mineral sector has been greatly reformed since the financial year 1997/98 by abolishing and streamlining some of the taxes. However, there are still problems of non-compliance, cheating and under-declaration, tax-evasion and smuggling of minerals to neighbouring countries. This has resulted in lower actual tax collections. The driving force behind these problems include the attitude of tax collectors, the tax rates and even more the absence of a well-defined mineral market.

There are a number of incentives given to investors in this sector, including tax holidays. These incentives have been criticised for discriminating against old investments, against domestic investors (entry qualifications) and concentration of capital (within the grace period).

Mineral production and exports have generally shown an upward trend. Likewise, the share of mineral production in GDP and in total exports has also been rising.

Among the factors behind production and export performance include: small scale miners' production increase after the 1980s liberalisation of the mineral sector (capital inflows, increase of small-scale miners).

The expansion of mining activities is associated with problems of security of tenure and other conflicts in the mineral sector. In terms of acquisition of mineral rights, the following developments have taken place:

a) A simplified process of acquiring mineral rights (after 1997/98);

b) Shorter time taken in acquiring licence;

c) Following the recent policy and legislative changes, the number of mineral rights issued increased by more than 2,000% from 11 in 1990 to 243 in 1999, although most of them are for prospecting;

d) Disputes related to encroachment were noted at Mererani, Arusha. This is because the size of claim holdings at Mererani was below the recommended size e.g. 300 square metres vs 50 square metres or 10 square metres. As a result, congestion, injuries and underground encroachment are common. Disputes between mining companies and claim holders centre around the system of allocating plots. This is particularly so in Geita.

e) Another set of disputes is between claim holders and mining workers (Village Communities). These disputes centre around job insecurity and conflict with *Meremeta's* programmes to provide equipment to small-scale miners, e.g. water pumps, compressors. Large-scale prospecting companies who enter into agreements with claim holders are worried at *Meremeta's* programme of supporting artisanal miners. The large prospecting mining companies tend to impose conditions that force claim holders not to authorise *Meremeta* to operate in their areas. The problem is acute in Mugusu, where an entire community feels very insecure about its future life, which signals unsustainable development.

f) Another category of problems is related to land disputes. The 1979 Mining Act, as amended in 1998, states explicitly how compensation with regard to farms, trees, graveyards, etc, is to be taken into consideration. The problem is poor implementation of this Act at the local level, resulting in numerous disputes which threaten social cohesion and harmonious relations.

The study has come up with a number of issues for policy consideration. These include tax holidays; the capacity to monitor and verify that mining operations comply with set regulations, mining in protected area networks ignoring EIA recommendations, e.g. Serengeti National Park, Mugusu Forest Reserve, Mkomazi Game Reserve; conflict between public and private interests and how to harmonise them; the importance of building a strong lapidary industry to add value to minerals.

Environmental Issues Arising include the following:

a) Growing number of artisanal/small-scale miners driven by the search for alternative income-earning opportunities. The period of economic reforms in Tanzania (since 1986) has witnessed movements of large numbers of people into mining areas, e.g. gold rush in Matabe Biharamulo, Rwamgasa, etc. With low education, lack of access to more efficient technologies and skills and insecure land tenure, the resulting in behaviour is detrimental to the environment.

b) Zonal offices are unequipped to monitor the rapid growth of mining activities. There are no mineral extension services. Policies to reduce government expenditure and restructuring of the civil service have resulted in retrenchment of staff in zonal mining offices thus affecting the capacity to monitor and enforce regulations.

c) Environmental problems observed in the fields require action by local authorities in collaboration with the National Environmental Management Council (NEMC). These include:

    i) Digging of thousands of mining pits, which are never refilled. This destroys the landscape.

    ii) Mining sites are abandoned as miners move to more lucrative sites.

    iii) Mining in river-beds leads to weakening of embankments, collapsing of banks, exacerbating erosion and siltation down-stream.

    iv) The location of mineral processing activities near rivers (e.g. sluicing and washing) has consequences during the rainy season when rivers swell and wash away tailings and remains of amalgamating chemicals such as mercury. When such materials are transported downstream, they affect other water users and change the habitat for flora and fauna.

    v) During the dry season, water from streams and rivers is recycled as much as possible through holding sumps. In the case of gold-mining, the danger is the subsequent accumulation of mercury sediments and its bio-magnification in the food chain within the aquatic ecosystem.

vi) Results from laboratory tests with respect to mercury concentration in human beings, surface water and soils are available from the Geology Department (UDSM) but they are still not being used to set standards. Cooperation is needed in this area. Indeed, there is limited co-ordination between authorities.

d) Disturbances in the biological environment, e.g. vegetation loss, contamination of marine organisms as well as other sensitive ecosystems and habitats.

i) Deforestation resulting from large concentrations of people in mining areas. People need wood for cooking, construction, timbering as well as clearing vegetation to create farms. Big mining companies practise so-called grid-clearing e.g. AAPS at Umba.

ii) Mining is taking place in forest reserves e.g. Mugusu, in national parks such as Ikorongo in Serengeti, in game reserves e.g. Mkomazi Game Reserve. The issue here is that there is no coordination between the relevant authorities.

e) Prevalence of environmental hazards and diseases to which people in mining communities are exposed. Examples include exposure to graphite powder and tailings in Mererani resulting in lung/respiratory diseases. In Mugusu a strange disease of swelling of the testicles and limbs is attacking the community, yet there is limited medical treatment. Accidents caused by underground blasting, falling rocks, inadequate ventilation and inhalation of mercury vapours and silicon dust could be reduced by observing regulations. Use of mercury retorts is extremely limited. This exacerbates the problem of mercury contamination in addition to adding extra cost to miners by inefficient use of the chemical.

Social Issues Arising include the following:

a) Unplanned movements of people into mining areas are the root cause of social problems. Gold rushes result in springing up of informal villages which lack basic sanitary facilities and which often bring problems of law and order. The ability to predict, plan for and provide basic services is lacking on the part of authorities at village, ward, division and district levels.

b) There are problems of diseases and inadequate medical services to deal with them. Also awareness and information regarding diseases and hazards is lacking. HIV/AIDS and STDs are common in mining areas. A care-free attitude is prevailing. The mentality of "*work, earn and enjoy*" is now common among people in mining communities.

c) There are areas of serious tension between claim holders and mine workers in terms of respecting contractual obligations and meeting basic needs. Contracts are in most cases verbal, informal, tending to disadvantage mine workers, who do have no organisation to safeguard their interests. Claim holders claim to represent the interests of mine workers.

d) General lack of awareness and information about mining policies, procedures and rules and regulations is prevailing. It is also a source of conflict and tension, e.g. at Rwamgasa there is a land-use conflict between mining and agriculture.

e) Problems of drug abuse are also common in mining communities (particularly bhang smoking and alcoholism). When miners get their money they spend it lavishly, thus encouraging prostitution as well. Women who do not engage in gainful activities within mining communities exacerbate the problem of prostitution. Attempts by religious bodies to provide a spiritual and counseling service are commendable but inadequate. Mining activities are also associated with strange and anti-social beliefs, e.g. engaging in rape to gain wealth; impregnating young girls and abandoning them, etc.

f) There is still a problem of child labour, although efforts are being exerted to prohibit it. School children at Mererani do not attend school regularly as they also engage in mining activities. In Kigwasi for example, there is no school, while in Kalalani, attendance is very poor.

This study has shown that macro-economic reforms that began in earnest in 1986 have stimulated the expansion of activities in the mining sector in Tanzania. There are more small-scale miners, mining companies and auxiliary activities today than a decade ago. New discoveries of mineral deposits are increasingly being reported. The policy response by economic agents in the mineral sector has subsequently impacted on the economic performance, the social fabric as well as the environment. For example, there are benefits in terms of capital inflows employment creation, improvement of infrastructure and the transfer of technology. All these have meant generation of income for individuals, companies and the government. The government has earned and continues to earn considerable revenues in the form of taxes, royalties, etc. There is no doubt that investors stand to gain through dividends.

These benefits notwithstanding, the study cautions that there are still many loose ends, which encourage leakage in revenues. These problems require various policy responses and implementation of existing rules and regulations. They call for raising the capacity of the ministry at central and zone level to administer and monitor existing laws, rules, regulations and procedures governing the mineral sector, as benefits accruing now fall short of their real potential.

Furthermore, the study findings sound a note of caution regarding the social and environmental costs arising out of expanded mineral activities during this period. A number of social problems resulting from unplanned movements of people into mining settlements are exerting a toll on the health of men, women and children who flock into these areas in search of income. These include the use of drugs, child labour and prostitution, all of which require immediate policy responses in terms of increasing the capacity to predict, plan for and provide basic services for mining communities. Environmental consequences of expanded mining activities, especially by artisanal miners, are becoming evident. Lack of modern technology and skills among such mine workers leads to environmentally-unfriendly mining methods e.g. digging thousands of pits and abandoning them. Loss of vegetation in surrounding areas results from increased consumption of wood products by large numbers of people in mining settlements. Unsafe use of chemicals, such as mercury, and mining in protected areas are issues that require institutional co-ordination and compliance of guidelines for conducting EIA as well as mitigation measures.

Mining development in Tanzania is not being performed in a manner that facilitates sustainable development. Unless prevention and mitigative measures are put in place, the outcome may be that of loss for the economy, the environment and society both today and in the future.

1

MINING DEVELOPMENT IN TANZANIA

Preamble

The economic history of any country is characterised by a series of policy and political changes that shape not only the economy *per se* but also the social structure and the physical environment. However, this is not a one-way causal effect. The state of the physical environment and changes therein also condition and shape, to a great extent, the possibilities of political and social manoeuvring of a given society as it strives to reproduce itself over time.

This study attempts to link market-based macroeconomic reforms in Tanzania with the positive and negative effects on the socio-cultural fabric of society and the natural environment. The mining sector is taken as a case study to illustrate this linkage and the sustainability implications arising therefrom. During the early years of independence (1961-1967), the government did not pursue economic policies that were significantly different from the last phase of the British Colonial Government. An attempt was made after 1967 to institute state control by nationalising the "commanding heights" of the economy (mining included) and to promote a collective-style rural development policy popularly known as *Ujamaa*.

During this period, an impressive record of social gains in terms of literacy, access to health facilities, water and political organisation was registered. However, at the same time, the foundations of economic performance were severely shaken.

By 1975, Tanzania was already experiencing economic difficulties which culminated in the period of economic crisis in the mid 1980s. Attempts to reform economic policies began with rather unsuccessful "home grown" programmes in 1981, a situation which led to the adoption of the IMF-inspired economic recovery programme in 1986. The case study of mining seeks to identify and establish links between the economy, society and the environment in relation to macroeconomic policies.

The Post-Independence Overview of the Mining Sector

Tanzania has diverse mineral resources.[1] The list ranges from precious metals, namely gold, silver and platinoid-based (and other) metals such as zinc, lead, copper, nickel, tin tungsten, cobalt and beach sand minerals. Others in the list include ferrous metals, namely, iron and chromium, precious stones and gemstones, i.e diamonds, rubies, tanzanite, gem garnets, emeralds, tourmaline, spinel, zircon, sapphire and peridot.Also found are industrial minerals namely, graphite, mica, kaolin, bauxite, etc. Energy minerals include coal, gas, uranium and petroleum. Evaporites include salt, soda ash, gypsum, limestone, etc.

Prior to independence in 1961 mining activities had become significant during the German colonial period (1884-1917) with keen interest in gold, mica, garnet, coal and uranium. Thus during German rule, mining was entirely under private ownership and/or in the hands of companies.

British rule (1918-61) was dominated by intensification of mineral prospecting and mining, with both small-scale and large-scale miners participating. There were South African Companies as well as British Companies operating in different areas in the country. Gold production started to do so with the advent of World War II and continued to decline with the discovery of diamonds at Mwadui in Shinyanga in 1939 by Dr. Williamson.[2] Diamond exports accounted for the largest component of export value while gold production kept on declining mainly because, during World War II, mining companies were denied priority in supplies

[1] A detailed listing of the minerals available in the country, production and marketing history is found in Chachage (1995). See also Nanyaro (1994)

[2] In actual fact diamonds had already been discovered by the time Williamson "discovered" them (Chachage 1995)

of machinery until the end of the war. Private ownership continued to dominate the mineral sector.

The Tanzanian economic and political policy evolution during the post-independence period can be categorised into three distinct but closely interrelated phases. The first phase consists of the early years of reconstruction, running from 1960 to 1967. Since Tanzania was more of a colonial-based market economy, mining was almost entirely in private foreign hands.

The second phase saw Tanzania adopting a central planning system. State monopoly was declared in 1967 through nationalisation of key sectors of the economy and subsequently the government acquired majority shares in mining companies and restricted foreign ownership of property. The State Mining Corporation (STAMICO) was established in 1972 to operate the mining sector in Tanzania.

STAMICO took over the Buck Reef Mine in Geita in the early 1970s and attempted to stimulate production of gold from alluvial sources by using large-scale production technology. Chachage (1995) observes that production of gold did not begin until 1981 and lasted only until 1989. Official data show that the eight years' production period yielded only 800 kilogrammes of gold. This was a large-scale operation, which needed massive injection of capital from the government, which was not possible following the inability of the government to improve its fiscal performance.

The third economic phase began in 1981. The efforts made by STAMICO to attract large amounts of capital and/or investments in the 1970s and early 1980s in the form of aid or joint venture also failed despite the fact that the new government mining law was in force by 1979. Through government Acts such as The Mining Act of 1979, the Petroleum (Exploration and Production) Act of 1980 and the National Investment (Promotion and Protection), it was no longer the preserve of the government to participate in such ventures and own majority shares.

As early as 1983 the government had formulated a mining policy geared towards small-scale mining and/or artisanal mining where by local small-scale miners could legally peg claims and work on them.[3] Although

[3] Classification on the distinction between Small-Scale Mining and Artisanal Mining is given in Appendix 1.

independent gold-mining companies began to appear in 1984 on the basis of the framework set out in the 1979 Mining Act, the beginning of serious liberalisation began in 1986 when artisanal mining started to become a major contributor to the national economy. The 1979 Mining Act was amended in 1982/83 and its framework was liberalised further during 1987/89 when a gold and gemstone trade rationalisation policy allowing for privatisation of the mineral trade came into force. Related to this new policy, the Bank of Tanzania also entered the gold buying markets on the basis of offering world market prices.

By 1992, mining activities (production and marketing) were liberalised and the free market was already in operation. As noted earlier, the liberalised mining sub-sector has not only attracted a good number of foreign investors but has also led to a dramatic increased participation of artisanal and small-scale miners who are basically motivated by positive liberal government policy and good prices offered by relevant banks.

The number of both small-scale as well as large-scale miners has increased tremendously during the period of liberalisation. Tanzania is expecting to produce a total of 1.3 million ounces of gold per annum (about 40 tonnes) following entry of the large-scale mining companies.

The inflow of foreign investment in mining has led to discoveries of many new gold reserves, thus production is expected to increase even further in the future. See also the Economic Survey 2001 (URT 2002). The future production increase was expected to come from, among others, the Golden Ridge at Lusu in Nzega district, Ashanti Goldfields, which runs a 50-50 joint venture with Anglo-gold of South Africa. This company operates a large open-cast gold mine at Geita, on the southern shores of lake Victoria. The mine had a planned annual production capacity of 500,000 ounces of gold (about 15.5 tonnes) from the gold reserves currently estimated to be 12.1 million ounces. The investment capital of this mine which will also be the second largest gold mining operation in Africa outside South Africa was expected to be US $ 165 million. This mine started production three months ahead of schedule and was inaugurated in early August, 2000.

The future production increase is expected to come from, among others, the Golden Ridge at Lusu in Nzega district. Ashanti Goldfields, which runs a joint venture with an Australian Company, Resolute Limited, completed construction of the mining site and commenced production in November 1998. It is estimated that the company has discovered a total of 2.4 million ounces of gold reserves and the target is to produce 180,000 ounces every year.

Bulyankulu gold deposits, owned jointly by Kahama Mining Corporation and Birrick of Canada, has the largest gold deposits in East Africa containing some 10.4 million ounces of gold. The company which came on stream in 2001 was expected to produce an average of 400,000 ounces per annum. In an effort to improve the infrastructure the company plans to construct a 75 kilometers road and a 50 kilometer water pipe from Lake Victoria.

A joint venture, between Randgold Resources of South Africa and Pangea Minerals of Canada, was issued a prospecting licence and mining licence (gold) at Nyalingo golden ridge in Shinyanga. The results show that the company has 2.2 million ounces of gold reserves. Production was expected to start in 2002 at an annual rate of 60,000 ounces.

Likewise Pangea Minerals in collaboration with Anglo-American of South Africa, are still prospecting at Mwime (Chocolate Reef). This area is estimated to have 2 million ounces of gold.

Also in the list is the East African Gold Mines Company at Nyabirama and Nyabigena areas in Tarime district. With a gold reserve of more than 4.1 million ounces, the company started production in 2001 at an estimated annual production of 180,000 ounces and investment capital of US $ 90 million.

Other mining companies that are still prospecting in Geita district include the East Africa Mines (Australia). This company is prospecting in Buziba and Rwamgasa in the area that used to be owned by Buckreef. Preliminary results indicate that 200,000 ounces per annum is expected to be produced.

As regards small-scale investors, the government recognises the positive contribution of the artisanal and small-scale mining sub-sector to the economy according to the Tanzania Mineral Policy (URT, 1997b). This contribution includes the discovery of mineral deposits, mineral production and the generation of income in the poor rural communities.

In view of the importance of the small-scale mining sub-sector, the government is committed to facilitating transformation of the present artisanal mining activities into more organised and modernised small-scale mining enterprises. Also the establishment and promotion of an organised mineral marketing network would be encouraged.

About 80% of the 1992/93 mineral revenue was from mineral production by artisanal miners while the remaining 20% came from both medium as well as large-scale miners (Nanyaro, 1994). The current proportions are likely to be different in that the contribution of medium and large-scale miners might have risen following the massive influx of large-scale investments in the mineral sector from 1994. This reversal is due to sophisticated and superior technology, which the mining companies use, compared with the inferior technology used by small-scale miners. At present, mineral production from medium and large-scale miners will still be lagging behind the output from artisanal miners, due to the fact that most large-scale investors are still at the prospecting stage.

Although financial sector reforms undertaken since 1991 have attracted new banks to Tanzania, the financial services available are still not tailored towards meeting the requirements of the mineral sector (URT, 1997b). This is because mining operations require huge capital outlays, which calls for the expansion and improvement of the financial intermediaries. Artisanal and small-scale mining operations require relatively small capital resources but the financial system has been slow, if not reluctant, to respond to the capital demands of this important sector. Most small-scale miners cannot meet the loan conditions of the formal financial sector, and because the local financial institutions do not fully realise the economic potential of the mining sub-sector they are almost reluctant to formulate credit schemes to support them.

Capital constraint is perhaps not the only obstacle facing the mineral industry. At present, there is a wide gap between the planners at central government and the remaining segment of stakeholders in the mineral sector. The Ministry of Energy and Minerals, which is charged with the responsibility of formulating the National Mineral Policy, does not seem to adequately involve other players in the mining sector in the policy nor in the strategies formulation processes Thus, the policy document is seen to be a product of the Ministry's technical staff who cannot be the true representative of all the interest groups despite their professional competence.

For instance, the Federation of Miners Associations in Tanzania (FEMATA), which is the policy advocacy forum for small-scale miners in the country, has repeatedly appealed to the Minister for Energy and Minerals regarding the irregularities in enacting the 1997 Mining Act. Over a long time this forum has been ardently making proposals and/or suggestions to the ministry as its contribution and way of participating in the decision and policy-making process without appreciable attention from the government. In November 1996, it was reported that FEMATA, which used to represent no less than 500,000 small-scale miners in the country, was by then completely excluded during the review and revision of the legal regulatory and fiscal regime for the mineral sector and the drafting of the Mining Act of 1996. A list of stakeholders who were fully involved includes Tanzania Chambers of Mines (whose constitution does not say much about the interests of the small-scale Tanzanian miners), foreign mining companies such as De Beers, Kahama Mining Corporation, Graphtan and foreign and local mining experts. FEMATA protested to the Minister for Minerals and Energy. Despite a verbal recognition by the ministry, FEMATA has been marginalised almost throughout the 1998 Mining Act formulation and/or preparation process. After realising that it has been deliberately ignored, FEMATA had to take other expensive measures to alert and arm parliamentarians (MPs), particularly those from the opposition parties. They were provided with adequate knowledge on controversial sections of the new Mining Act so that they (MPs) could raise pertinent issues suggested by FEMATA before the endorsement of the Act by Parliament. According to FEMATA, this approach has been extremely effective with one of the FEMATA top executives acknowledging the government's revised policy and measures as presented in the 1998 Mining Act.

> "I am not disputing what others have said, but let me say we are very proud and gratified that the 1998 Mining Act has accommodated most of the fundamental policy issues pertinent to the small-scale miners. The satisfaction is particularly great stemming from the fact that it has taken us very long to get our views into the government's corridors."

At this stage one should note that, for effective implementation of policies, the government needs to work in collaboration with its partners by way of involving them fully in the strategy and/or policy formulation process. Government partners are actually other stakeholders who are also key players in the mineral sector and whose consent and support is particularly important if sustainable development is the objective. The legal framework (legislation), which is mentioned in the Mineral Policy document as one of the focus areas, is likely to falter if the mineral legislation formulation procedures do not adequately involve the stakeholders.

The development strategy according to the 1997 Mineral Sector Policy, places strong emphasis on private sector participation in the mineral sector, and reduces the government's direct involvement in mineral production activities (see also ESRF 1997). The government will only continue to play the role of regulator, promoter, facilitator and service provider. An immediate question concerns the capacity of the country to adequately perform these important duties. Tanzania is among the many countries in Sub-Saharan Africa which are still in the infant stage in the struggle to develop their capacity (capacity building) to manage the economy efficiently. It may therefore be too early for the government to assume all these demanding roles.

By way of concluding this section, it can be said that there appears to be a strong relationship between macroeconomic reforms and the growth of productive activities in the mining sector. The natural environment has drastically been degraded as a result of the mining activities despite the fact that good rules and regulations are in place. The loss of revenue which the government fails to capture, because of its inability to administer tax collection from small-scale miners, is enormous. Other reasons include policies and programmes pursued by the government and developments in the mining sector. The decline in mineral production experienced during two decades (1967-1986) can be attributed to the government's attitude and approach towards private investment (both local and foreign), strong restrictions and controls in all spheres of production and distribution, cumbersome regulatory procedures (bureaucracy), unattractive taxation arrangements and the unstable macroeconomic environment.[4]

[4] See Nanyaro, 1994

On the other hand, the recent positive performance in the mineral sector seems to be associated with the liberalised economy in general and the mineral sector in particular. Likewise, the macroeconomic policy regime has encouraged the participation of small-scale miners (SSM) despite their activities being associated with a score of adverse environmental and social impacts.

Environmental and Social Outcomes of Economic Reforms in the Mining Sector: An Overview

Although this study is focusing on the mining sector, because of its relationship with other sectors it is necessary to address the problems of the mining sector in association with these other sectors. For example, the expanding mining sector draws financial resources from financial institutions as well as labour resources from the agricultural sector. Thus the environmental and social impacts of the restructured mining sector are likely to spill over into other sectors such as agriculture.

It was noted earlier that in its recent pronouncements, the government has reiterated its resolve to disengage from directly productive activities. However, it has pledged to continue its role of ensuring an enabling policy environment and the provision of a conducive economic infrastructure for the main economic sectors of the economy (URT, 1996; URT, 1997b). The government's policy for the mineral sector places emphasis on attracting and enabling the private sector to take a lead in exploration, mining development and marketing.

Under the reform package, the role of the government sector is to stimulate and guide private mining investment by administering, regulating and promoting the growth of that sector. This includes regulating and transforming artisanal mining, facilitating efforts towards the reduction and/or elimination of adverse social and environmental impacts and alleviation of poverty especially for small-scale miners. Some environmental and social impacts associated with mining activities are beginning to be acknowledged. These impacts can be grouped into the following categories:

- Short, medium and long-term
- Direct and indirect

- Reversible or irreversible
- Local or regional (global)
- On-site or off-site
- Beneficial and adverse

Environmental Impacts

Mining operations can be a source of environmental destruction if they are not well managed and controlled. Environmental impacts of the production of minerals depend on such factors as the ecological character of the mining site, the quantity of materials being removed and the chemical composition of the surrounding rocks and soils as well as technologies being used (Chachage, 1995a). Mining has an impact on the physical and chemical characteristics of the environment, e.g. soil, water, atmosphere and other physical resources, on top of its influence on biological entities, e.g. flora, fauna, their structure and functions.

Most concerns for the environment are normally directed at mining operations, usually referred to as metal mining (URT, 1997a). However, the on-site and off-site impacts due to quarrying and non-ferrous mining operations can also be extensive. Destruction of natural habitats at the mining site and at waste disposal sites can take place if mitigation measures are not adequately undertaken. The influx of miners and other people may result to encroachment as well as destruction of adjacent habitats. There are cases where mining operations cause land degradation due to inadequate rehabilitation after closure. Soil contamination resulting from untreated residues and spillage of chemicals is also common. Drainage from mining sites from leaching processes is also a source of pollution as well as that resulting from oil and fuel spills, air and dust emissions from mineral processing activities. A case study by Chachage (1995a) shows that displacement of wild animals, deforestation, pollution and land degradation are the price Tanzania has to pay as a result of mining operations especially in Mahenge District.

Economic, Social and Cultural Impacts

Apart from the negative impacts mining has on the natural environment, mining activities are also known to change cultural factors, e.g. land use, recreation, aesthetic, cultural status, man-made facilities and activities and other human interests. It also accelerates health hazards. In most cases the changes are adverse.

a) *Employment*

Restructuring of the mining sector has led to an expansion of mining activities in the country, which has also resulted in increased employment in the sector (Nanyaro, 1994). In early 1990s, over 80% of the mineral revenues were derived from mineral production by artisanal miners who had not only been increasing in number but also becoming active participants in the mining industry. It is estimated that during this period nearly 550,000 people were directly engaged in artisanal mining alone (Mutagwaba, 1997). By supporting the activities of small-scale miners, who are among the 80% living in rural Tanzania, their social and economic conditions will gradually improve, thus contributing to reducing rural poverty and unplanned migration to towns and cities.

Expansion of mining activities by small-scale miners in recent years has been taking place at old mines and areas where known alluvial deposits existed up to the early 1970s. Mining activities experienced a spectacular boom in 1976 with the discovery of new deposits. Almost every year, small-scale miners continue to make new discoveries in many parts of the country. Following these discoveries, reportedly about 500 gold mining sites and approximately 300 gemstone mining sites were being operated by small-scale miners by 1992 (Chachage, 1995a). Bulyankulu deposits in Shinyanga Region and Mpepo deposits in Mbinga District are examples of such new discoveries. Production of gold at Bulyankulu is planned to start in 2001. Lusu area located in Nzega District is another place where discoveries of gold reserves have been made. The project has resources of approximately 2.8 million ounces (nearly 86 tonnes) of gold with the potential of finding additional reserves in the immediate neighbourhood.

Expansion of mining activities tends to draw economic resources away from less rewarding sectors. For example, since miners have been motivated by positive, liberal government policy and attractive prices offered by designated buyers, the mining sector has tended to draw labour away from the agricultural sector which can improve the per capita incomes in the sector particularly in areas where there is surplus labour. However, while employment in the mining sector may increase, the actual change of employment in the agricultural sector will depend on whether or not there is surplus labour. In a situation of agricultural labour shortage, withdrawal of labour from agriculture could actually exacerbate the situation. Since mining activities are beginning to be attractive, financial resources also tend to flow towards the mining sector. By withdrawing resources from less rewarding sectors of the economy, employment opportunities in those sectors may consequently shrink.

b) *Tenure Issues and Displacement of the Local Population*

It is clear that the mining industry in Tanzania is developing rapidly and a number of potential investments continue to flow into the sector. Most of these will probably be large open pit operations. Expansion of such mining operations is known to dislocate the local people who are then normally forced to migrate and start new settlements elsewhere. Local people will therefore be shifted to other areas away from traditional lands; their cattle will miss the rich grazing land and children will need alternative school premises and facilities. The experience of the 1970s settlement schemes in Tanzania shows that social tensions and conflicts with environmental depletion are the obvious negative consequences of new settlements. It is important to ensure that resettlement of the population due to expansion of mining activities does not repeat past mistakes. Other conflicts are related to land, wildlife and water resources.

The procedures for granting investors mining rights is clearly stated in the National Mineral Policy of 1983 and the 1998 Mining Act. However, the problem is that actual practice has tended to disrupt and displace not only the local people, but also other economic activities in favour of the mining operations. In many instances, customary law has found itself in conflict with other legal and/or government institutions especially due to the rationale of conferring ownership of mineral rights onto investors without due consideration of local customs and values and without the general involvement of the indigenous people.

c) *Prostitution, Crime Rates and Erosion of Ethics*

Social problems such as prostitution and sexually transmitted diseases (STDs) have tended to increase within the mining areas. Small-scale workers are mostly men from distant areas of Tanzania. They seldom move to the mining sites with their families and work for long periods without going back home. In such circumstances, acts of sexual misconduct cannot be ruled out. Consequently, sexually transmitted diseases have been on the increase in Mahenge District for example, coinciding with the intensification of mining activities (Chachage, 1995a). Following the expansion of mining activities at Mpepo in Mbinga District, a wave of crime took residents there by surprise. In addition, there have been a number of accidents in some of the mines where many Tanzanians were injured or lost their lives. A serious one was the April 1998 accident at Mererani, in which over 100 people were killed. Since then, many other fatal accidents have occurred yet claiming lives of young and productive Tanzanians. The last two occurred between June and December, 2002 claiming a total of 2 and 9 lives respectively.

According to a local Non-Governmental Organisation (NGO) based in Babati District (Arusha Region), The Human Rights Defender, there are serious violations of human rights in Babati District, caused by the belief that if unlawful sexual acts are committed, the individual becomes more successful in terms of mineral harvesting. There are cases where people are forced to rape young girls, mostly those who are under age (following the advice of traditional healers or traditional medicine men), so that they become fortunate in mining operations. Others are said to force temporary marriages with young girls who are subsequently abandoned immediately after they have given birth.

Evidence from Mahenge has shown that some of the foreign investors become unnecessarily aggressive in their operations. Domestic small-scale investors are in many instances dominated by foreigners to the point of crowding out the local ones from claim holdings, thus depriving them of their source of income.

d) *Health Hazards*

The rapid spread of sexually transmitted diseases and pollution are not the only health hazards associated with mining operations. The influx of a large number of people into mining sites where facilities are inadequate causes health problems as well as a breakdown of law and order. Many miners are exposed to hazardous chemicals used to process minerals. A government report, URT (1997), mentions asbestos, cyanide, mercury or other toxic materials used on sites as some of the dangerous chemical substances.

Small-scale mine workers are insufficiently financed and lack the necessary professional knowledge. As a result they operate without adequate working gear. Lack of protective clothing and other working equipment suggests that small-scale miners are risking their health due to exposure to occupational hazards such as accidents, heat, noise, vibrations and dust inhalation. Similarly, unsanitary living conditions and malnutrition among the children whose parents are working in the mines are said to be on the increase. Such a state of affairs threatens the future status of the national work force. Unchecked health hazards may reduce the capability and/or productivity of a future work force.

Artisanal miners have increased their participation in the mining sector in recent years. The consequences of their activities are only now beginning to be known. Indeed, their activities do improve socio-economic conditions but they also lead to serious environmental consequences. The socio-economic potential pertains to employment creation, income generation for individuals and the government, while the destructive aspect relates to poor health and safety, land degradation, pollution of water bodies and air. Owing to the fact that large-scale miners take over the mining areas which used to be under small-scale miners, the relationship between the two groups has not been friendly.

Macroeconomic Reforms and the Mining Sector: An Overview

Macroeconomic Policy Objectives and Targets

Although the country's economic performance during the 1960s and part of the 1970s was to a large extent satisfactory, signs of difficulties were seen from 1973. The government responded to the emerging economic crisis by adopting economic stabilisation and reform policies after a long period of resisting calls from the IMF and the World Bank for a market-oriented economic system. Thus, since early 1980s Tanzania has implemented macroeconomic and sectoral policies to accelerate economic growth, which was seen to stagnate during the pre-reform period. The effects of these policies have been felt in the mining sector.

Macroeconomic policy history, economic reform programmes and the subsequent economic performances are widely documented. Recent scholarly reviews include the work by Lipumba and Mbelle (1993), Sankhayan (1994), Bagachwa, *et al* 1995, Putterman (1995) and Wangwe and Ndulu (1997). It is not thus our intention to replicate that information. It suffices to note that Tanzania has gone through four major economic reform programmes. Two of these, initiated from within, are the National Economic Survival Programme (NESP) in 1981 and Structural Adjustment Programme (SAP) of 1982. The other two programmes supported by the World Bank and IMF are the Economic Recovery Programmes (ERP I and ERP II).

ERP I was adopted in 1986 (to 1989) and was followed by a similar programme, the Economic and Social Action Programme (ESAP) or ERP II, implemented between 1989 and 1992. The ESAP was basically aimed at taking the path set by ERP I, i.e. continuation of the recovery process to achieve sustained growth and improved social welfare, apparently with the intention of cushioning the harsh impacts of ERP I on social services.

The main macroeconomic reform policy objectives as contained in ERP II as well as in the First and subsequent Rolling Plan and Forward Budget (RPFB) are meant to move away from a controlled economy which is dominated by the public sector. The intention is to achieve an economic system which is increasingly liberal in nature, much more market-oriented in its operation and dominated by the private sector. The new policies are generally designed to correct structural imbalances in the economy and spur investments. These objectives can be summarised as follows:

- To achieve and sustain macroeconomic stability;
- To liberalise trade and foreign exchange regimes;
- To create a strong and efficient private sector and reduce the government's direct involvement in productive activities;
- To improve efficiency in the public sector and especially in the use of public resources;

- To achieve a sustainable development path; and
- To combat poverty and improve the living standards of the people.

Along with these objectives are overall macroeconomic policy targets, which are by and large an extension of the targets highlighted in the Third Rolling Plan and Forward Budget. These targets include:

- Promotion of economic growth: The target was to achieve a real annual average growth rate of 6%.
- Reduction of inflation: The target was to attain 15% inflation rate by 1996/97, thereafter a single digit of 5% inflation rate was targeted for 1998/99.
- Enhancement of central government savings: A savings of 0.5% of GDP was projected for 1996/97, rising to 2.5% in 1998/99.
- Strengthening the country's balance of payments position: A decline in the external current account deficit (excluding grants) to 10.9% of the GDP was targeted for 1996/97, falling to 10.4% in 1998/99. Also, the gross official reserves to the equivalent of at least three months of imports are projected throughout.
- Maintenance of monetary stability and strengthening the financial sector: the target is to accelerate structural and institutional reforms in the financial sector.
- Continuation of the civil service and the parastatal sector reforms: the target is to continue with the reform programme addressing the reduction of the size of the public service, rationalisation of pay and incentives and reducing government functions to core ones. Parallel to that is the target of continuation of parastatal sector reforms focusing on restructuring and privatisation processes.
- The policy instruments deployed during the implementation process are described in the coming section.

The Economic Policy Instruments

In order to achieve the objectives and targets of the economic reform policies, a number of economic instruments have been deployed. The policy instruments are believed to have had varied but significant impacts on the mineral sector. The instruments used to achieve the above outlined objectives and targets include:

- Fiscal instruments such as export taxes, import tariffs, import quotas, government budget subsidies; etc.
- Monetary instruments such as, money supply, interest rates; etc.
- Price deregulation, trade and market liberalisation and diversification of public enterprises.

The effects of these policies are likely to be felt in many sectors in terms of benefits as well as costs.

Overall Impacts of the Reform Policies

The government has continued to follow ERP policies through its Rolling Plan and Forward Budgeting approach and the country's social and economic situation appears to be changing in that, while there are noticeable improvements in many areas, economic reforms have also created a long list of costs to Tanzanian society. Among the undesirable outcomes are deterioration in social services delivery, displacement of the local population, loss of property rights and environmental damage. Nevertheless, in other areas reform measures have produced some positive results as they have led to the recovery of growth, price stability and have put goods of every description onto Tanzanian shop shelves, though exorbitant in prices.

Tanzania of the 1990's was distinctly different from Tanzania of fifteen years earlier in many aspects. The performance of the economy prior to and after the economic reforms is presented in Table 1.1. A marked improvement in some indicators during the post reform period is evident. For example, the real GDP per capita has generally been growing during the period under investigation. Impressive performance of the economy was spearheaded by agricultural activities and exploitation of natural resources (Wangwe and Ndulu 1997). The agricultural sector registered a real annual growth rate of 5.7% during the reform period against a 3% growth rate during 1980-85. Inflation after 1986 has been averaging 26% which is marginally lower than an annual average of 31% prior to 1986. Inflation in 2002 stood at 4.1% (BOT 2002).

The real per capita income, which used to grow at an annual average rate of -5.8% prior to 1986 is now increasing at a positive rate (average for the 1990s).[5]

On the other hand, figures on real GDP per capita in dollar terms do not appear to reveal an upward after 1986. This is obviously the result of a fixed exchange rate regime, trend which was followed in the country prior to the reform period. The regime distorted the exchange rate and therefore the value of the Tanzanian shilling was such that the domestic currency was substantially overvalued. Devaluation of the Tanzanian shilling, which accelerated in the late 1980s, might in principle have expressed the real value of the local currency. Since 1982 a series of major devaluations of the Tanzanian shilling have been initiated. A 10% devaluation was effected in 1982, followed by a devaluation of 20% and 36% in June 1983 and June 1984, respectively. Then came the biggest devaluation in June 1986 after reaching an agreement with the IMF. This was followed by 23% and 29% devaluations in November 1988 and December 1989 respectively (see also Kaufmann and O'Connell 1991). Consequently, the Shilling, which stood at 7.1 to the dollar in 1970, reached 32.7 to the dollar in 1986; 99.3 in 1988, and 195 in 1990[6]. Autonomy in the allocation of credit and foreign exchange was granted to financial institutions when the Annual Finance and Credit Plan, the Annual Foreign Exchange Plan, foreign exchange controls and import controls were scrapped in 1991 and 1992 (Nyagetera and Tarimo 1997).

[5] See Table 1.1

[6] Putterman (1995) and Nyagetera *et al.* (1997)

Table 1.1: *Selected Macroeconomic Indicators: Tanzania 1980-1999*

Sn	*Economic Indicator*	*1980*	*1981*	*1982*	*1983*	*1984*	*1985*	*1986*	*1987*	*1988*	*1989*
1	GDP real growth %	-	-	-	-	-	-	-	-	-	2.7
2a	GDP per capita $ (current Prices)	245	262	295	280	253	299	205	137	203	170
2b	Real GDP per capita $ (1992 Prices)	-	-	-	-	-	-	-	-	-	345
2c	Rate of Inflation (%)	30.3	25.7	28.9	27.1	36.1	33.3	32.4	30	31.2	30.4
3.	Govt deficit/GDP (%)	-13.25	-14.42	-11.62	-9.72	-9.82	-8.76	-11.36	-8.55	-6.78	-7.03
4.	Exports/ Imports (%)	49.97	58.98	38	46	44	29	33	34	35	34
5.	Investment/ GDP (%)	29.52	27.12	23.61	17.37	17.64	16.9	23.82	23.84	22.4	21.7
6.	Exchange Rates	8.2	8.28	9.28	11.4	15.29	17.47	32.7	64.26	99.29	143.38
7.	Exchange Rate Change	-	1.0	12.1	22.8	34.1	14.3	87.2	96.5	54.5	44.4

Sn	*Economic Indicator*	*1990*	*1991*	*1992*	*1993*	*1994*	*1995*	*1996*	*1997*	*1998*
1	GDP real growth %	6.2	2.7	1.8	0.4	1.4	3.5	4.2	3.3	4.0
2a	GDP per capita $ (current Prices)	157	177	162	147	153	171	210	240	257
2b	Real GDP per capita $ (1992 Prices)	261	232	169.4	121.5	95.5	85.11	85.40	81.3	82.7
2c	Rate of Inflation (%)	35.9	28.8	21.9	25.2	33.1	28.4	21	16.1	12.8
3.	Govt deficit/GDP (%)	-7.66	-7.97	-4	-10.8	-7.9	-4.3	-1.5	-3.3	-4.6
4.	Exports/ Imports (%)	34	31	32	38	47	58	62	63	48
5.	Investment/GDP (%)	40.8	29.5	32.7	32.2	30.7	19.8	17.0	15.2	15.0
6.	Exchange Rates	195.06	219.16	297.71	405.27	509.63	574.76	579.98	612.12	606.75
7.	Exchange Rate Change	36	12.4	35.8	36.1	25.8	12.8	0.9	5.68	6.89

Sn	*Economic Indicator*	*1999*	*2000*	*2001*
1	GDP real growth %	4.7	4.9.	5.6
2a	GDP per capita $ (current Prices)	255	258	265
2b	Real GDP per capita $ (1992 Prices)	63.8	64.6	61
2c	Rate of Inflation (%)	7.8	6.0	5.2
3.	Govt deficit/GDP (%)	-4.1	0.3	7.1
4.	Exports/Imports (%)	52	53	46
5.	Investment/GDP (%)	15.5	19.1	18.3
6.	Exchange Rates	800	802	872
7.	Exchange Rate Change	0.3	8.7	9.4

n.a. = not available

Source: Shitundu and Mjema (1999) and Economic Survey (Various Issues).

Response of the Mining Sector

A step-by-step devaluation process described above was intended to open up and free the trade regime of the country for both external and domestic transactions. This is the period when private investors started investing in the mineral sector and private traders could freely sell minerals to appointed banks and designated dealers.

The exchange rate and interest rate regimes are now deregulated and the government has liberalised imports of capital goods used in mining activities.

The outcome of the application of the fiscal measures on the mining sector has been a boom in mining activities as evidenced by the prevalence of new investments through increased number and capacity of mining companies (large-scale miners), increasing number of new applications for mineral rights (prospecting licences, claim titles and mining licences). Table 1.2 gives the trend in the number of large-scale approved reconnaissance, prospecting and mining licences from 1990 to 1999.

Table 1.2 shows that the number of mineral rights (mining) granted in the country has been increasing, an indication that investments in the mineral sector have also been expanding. Likewise figures on mineral dealers (marketing) licences depict an increasing trend, which also confirms the assertion that, following the measures taken by the government including mobilisation of investors, mining activities have experienced a rapid expansion. For example, while only 11 licences were issued in 1990, a total of 262 and 413 were respectively approved in 1999 and 2001 which translates into a more than 2350% and 3750% increase between the two periods.[7]

Table 1.2: *Large-Scale Mineral Rights Granted from 1990 to 2002*

Sn.	Type of License	*1990*	*1991*	*1992*	*1993*	*1994*	*1995*	*1996*	*1997*
1	Reconnaisance	n.a	n.a	11	10	30	13	40	21
2	Prospecting	9	18	72	59	132	100	192	351
3	Mining	2	7	9	4	2	n.a	5	13
4	Mineral Dealer	n.a	n.a	n.a	n.a	139	157	n.a	n.a
Total		11	25	92	73	302	270	237	385

Sn.	Type of License	*1998*	*1999*	*2000*	*2001*	*2002**
1	Reconnaisance	5	23	n.a.	11	8
2	Prospecting	321	218	224	182	45
3	Mining	11	11	25	46	3
4	Mineral Dealer	n.a.	19	226	174	11
Total		389	262	475	413	67

* Denotes that data for 2002 are for January and february only.

Source: *Ministry of Energy and Minerals and URT: Economic Surveys (Various Issues)*

Growth of the mineral sector is also evidenced by production, mineral exports, employment and total value of investment in the sector which have revealed an upward trend particularly during the adjustment period.[8] Other notable expansions relate to the increasing number of mining towns and settlements (villages), expansion of auxiliary activities and increasing importance of the mineral sector to government coffers.

---

7 See also figures in appendices 4 and 5

8 These trends are discussed further in the empirical chapter.

In Arusha (one of the leading regions in gemstone production), the number of mining companies has nearly doubled, increasing by 43% from 58 in 1995 to 83 in 1998 (URT 1998e).

Objectives and the Working Hypotheses

Objectives of the Study

The main objective of this study is to analyse the macroeconomic policy reforms and the policy instruments employed in realising those objectives with a view to assessing the subsequent response of the key players in the mineral sector. This is followed by an investigation of the impacted areas in terms of economic benefits, social and environmental consequences. The intention is to establish whether or not the expanding mining activities can ensure sustainable development in Tanzania. If not, then the results should inform policy dialogue to come up with alternative policy options towards achieving that. A sustainable development path rests on three pillars, namely, economic sustainability, social sustainability and environmental sustainability, and it is defined as "development that meets the needs of the present without compromising the ability of future generations to meet their own needs" (WWF 1993). The concept of sustainable development emphasises broad socio-economic as well as environmental considerations.

At this stage it is important to bear in mind that renewable (non-exhaustible) resources are different from non-renewable (exhaustible) resources. Exhaustible resources include fossil fuel, gas, coal and non-energy minerals (Perman *et. al*, 1996). Thus mining cannot be sustainable because it is a depletable resource. Sustainability of mining activity is therefore gauged through reinvestment of the accrued benefits in other areas so as to stimulate economic growth and improve the welfare of the people.

Working Hypotheses

Six main hypotheses have been formulated to guide the analysis.

1. Liberalisation of the mining sector stimulates mining activities hence increasing the sectoral contribution to the National Income.
2. Lack of reliable marketing channels results in illegal exports and

therefore what is captured by official statistics is by far below the potential.

3. Movement of people into mining areas and the increasing number of activities result in more environmental degradation in the mining areas.
4. Lack of information and inability to participate in the market leads to large disparities in income among artisanal miners.
5. Mining development causes more negative than positive health and social impacts in the mining areas.
6. Inadequate involvement of local authorities in matters related to mining activities in their areas causes disputes and loss of property.

2

RESEARCH DESIGN AND METHODOLOGY

Study Areas

Three study areas were selected based on the importance of each in terms of types of minerals produced, social and economic relevance as well as policy significance (See Map 1 for mineral sites in Tanzania). Two of them are located in the northern part of Tanzania. These are Mererani in Arusha Region and Geita in Mwanza Region. The third research area is Umba Valley which is located in the north-eastern corner of Tanzania (see Map 2).

The three study areas are among the major producers of minerals e.g. tanzanite, gold and other gemstones. Both local and foreign investors have been extensively involved in mining operations in the above-mentioned mining centres during the reform period. In all three study sites there are interesting and controversial issues with regard to the growth of mining activities as they lead to growing social and environmental consequences some of which require policy responses. For example, Mererani produces tanzanite, most of which finds its way into Kenyan markets. This has the potential of reducing potential benefits to the country. Likewise, as pointed out already, in April 1998 a major accident occurred, which claimed the lives of over 100 people when mining pits were flooded with water and silt. This points to the issue of security and safety measures in the mining pits.

Also both in the Umba gemstone mines and in the Geita gold mines, rights of ownership have been granted to outsiders in a manner that has created ill-will among local people.

Another pertinent issue being explored is the rate at which the socio-economic problems are manifesting themselves and environmental problems are widening. At Mahenge in Morogoro for example, a Tanzanian miner was shot dead and two others were injured following a struggle for the mining rights between the local people and newcomers who happened to acquire the right of ownership. Also in Umba, guards of a foreign company killed two people. Such incidents create tension among the people living in the mining areas. Furthermore, rivers are being polluted with heavy metals like mercury. Vegetation is cleared, leaving the soil open to agents of soil erosion, while mine workers are exposed to hazardous chemicals. These issues require policy dialogue to ensure that unsustainable outcomes of expanding mining activities are reduced, if not eliminated (Maps of the study areas are shown in Maps 2 to 5).

Data Requirement and Sources

Both secondary and primary data sources have been utilised to obtain qualitative and quantitative data. Secondary data sources include published and unpublished materials from government ministries and departments responsible for mining both in Dar es Salaam, Dodoma and zonal offices in the study regions and districts. Also miners and mineral dealers' associations such as Federation of Miners Association in Tanzania (FEMATA), Tanzania Mineral Dealers Association (TAMIDA) and the Chamber of Mines were contacted. However, due to paucity of published information on mining activities and small-scale mining in particular, primary data were generated through surveys carried out in the three mining areas, to complement secondary data.

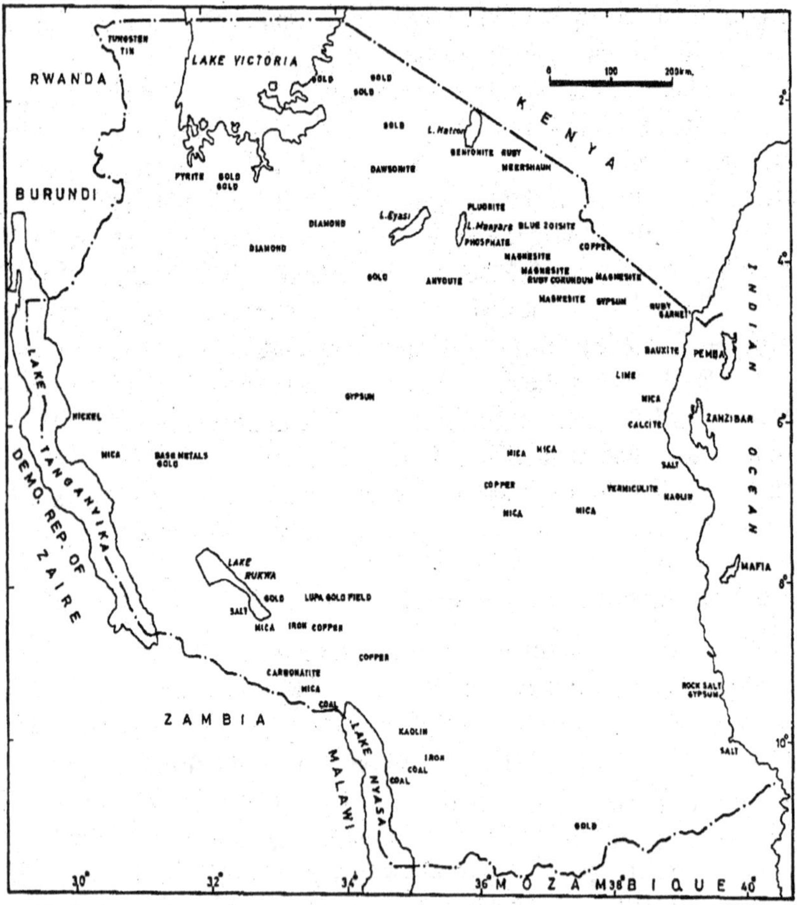

Map 1: *Tanzania Mining Sites and Potential*

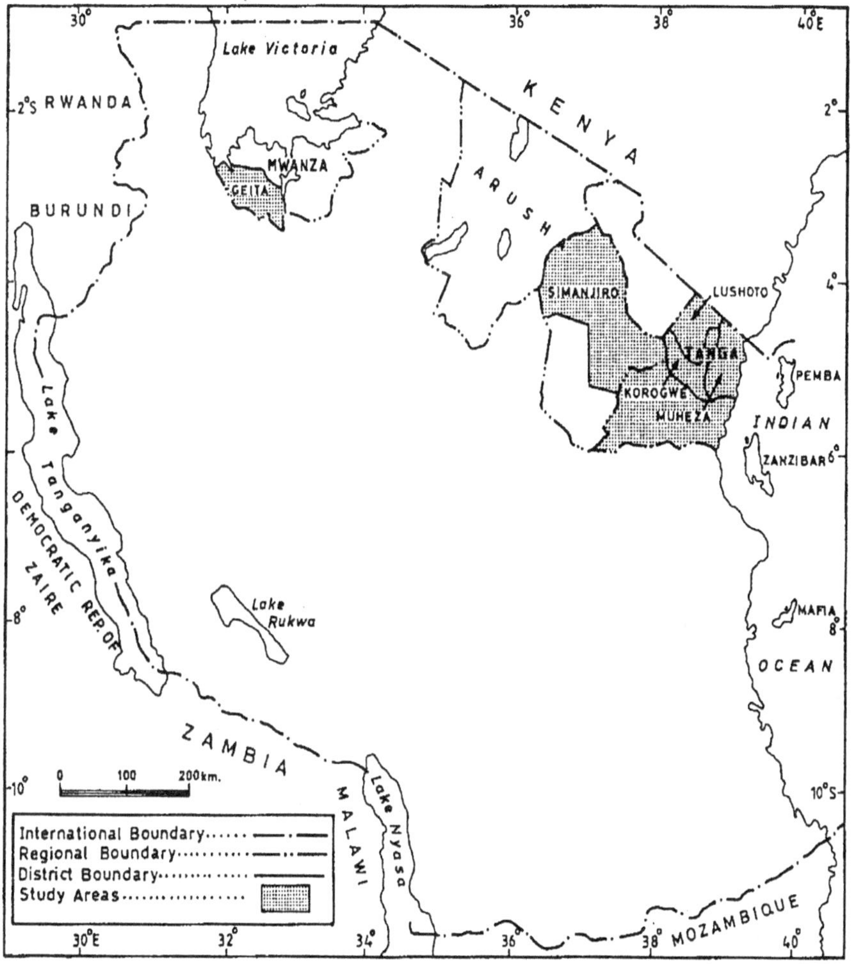

Map 2: *Location of Study Area: Arusha, Mwanza and Tanga*

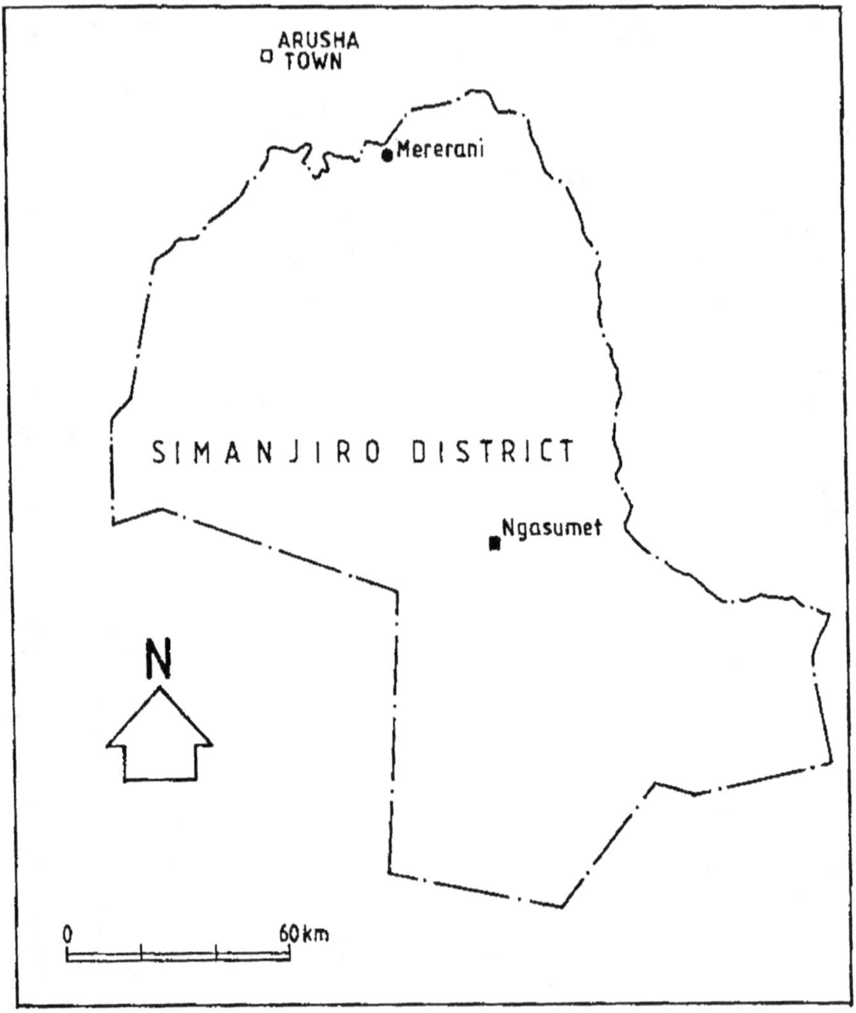

Map 3: *Simanjiro District Showing Mererani*

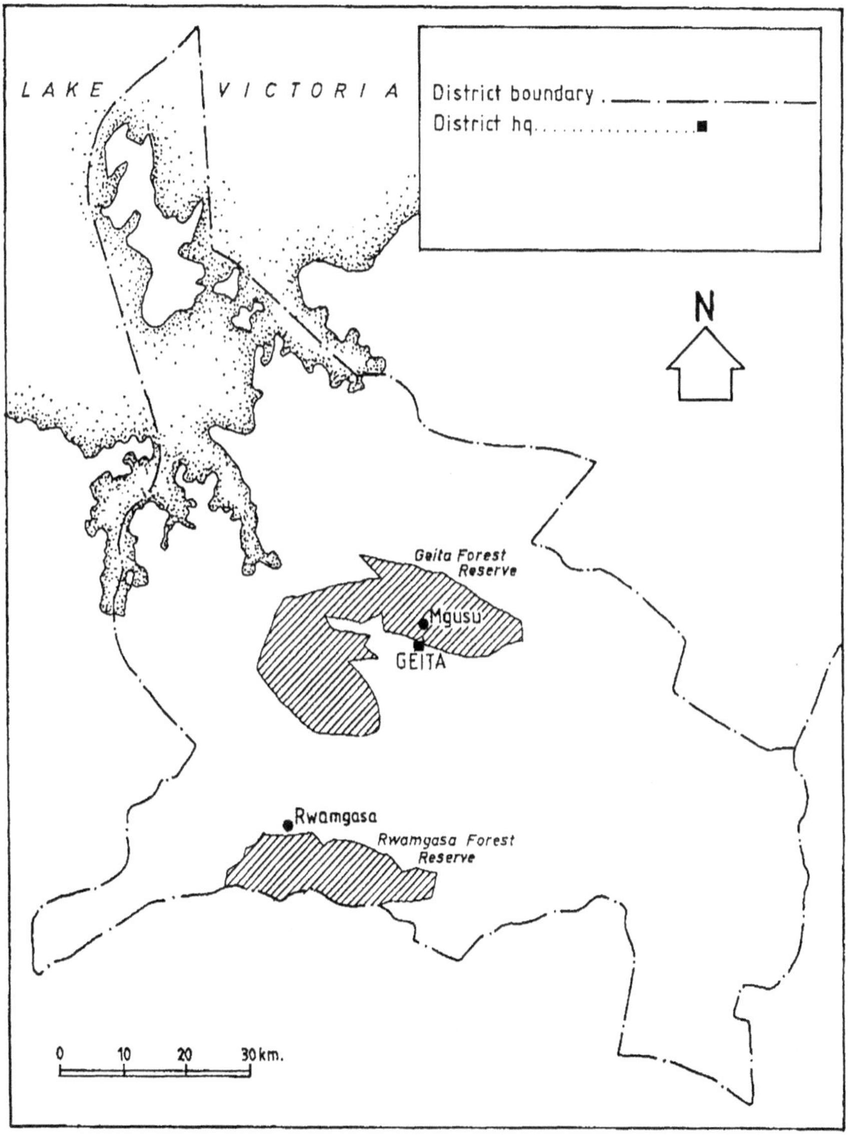

Map 4: *Geita District Showing Mgusu and Rwamgasa*

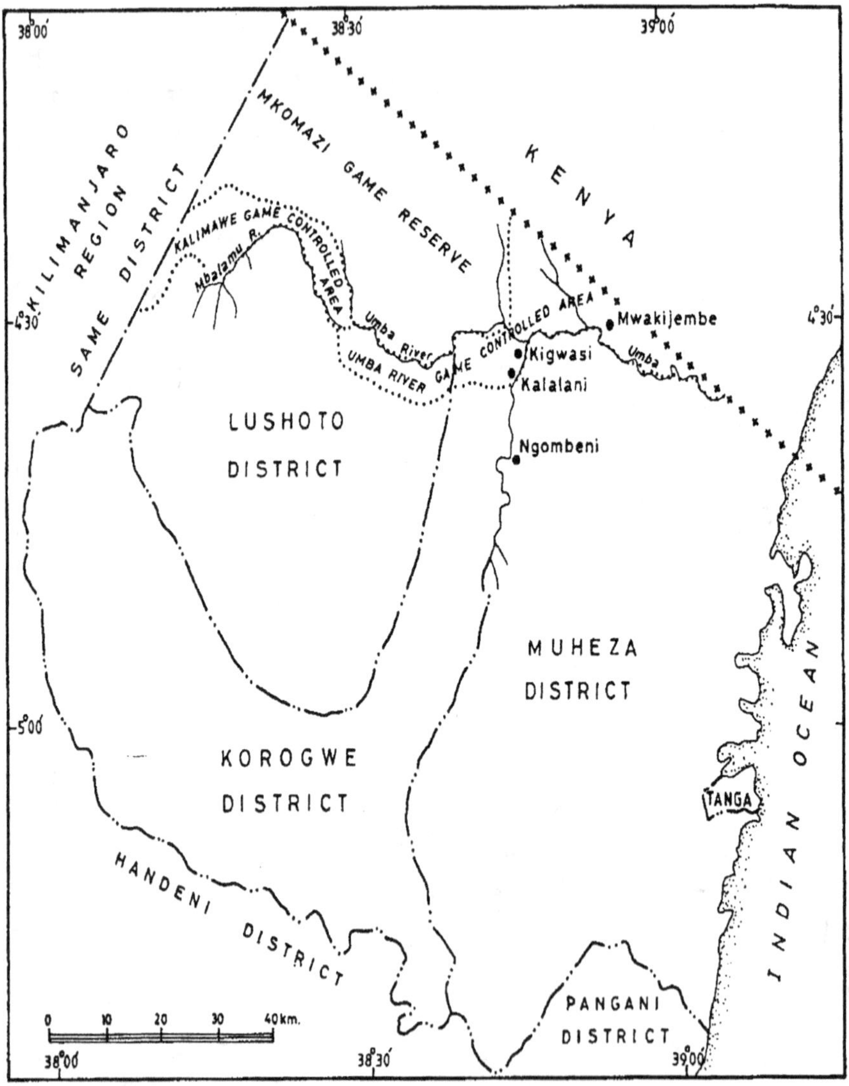

Map 5: *Korogwe, Lushoto and Muheza District Showing Kalalani, Kigwasi, Mwakijembe and Ng'ombeni*

Data Collection Methods

Field Approach

The approach in each area was to begin with the zonal mining office to get an overview of mining activities in the area served by the office and to get to know the people and agencies involved in mining. Thereafter, the research team went ahead to interview mineral dealers, mining companies, mining associations and other influential individuals in the mining sector. It was only then that the team moved to actual mining sites to talk to claim holders, mining workers and service providers. A structured questionnaire was administered to respondents, while two sets of interview guides were used to conduct unstructured interviews. Results from the unstructured interviews were recorded in note- books and compiled as raw field data[9].

Sampling Techniques

Sampling techniques included purposeful, stratified and random sampling to gain a wide range of activities of interest and within them to achieve some measure of representation. The choice of study area has been done deliberately in order to capture mining activities involving both local and foreign operators, small and large-scale mining activities.

Within each area, the target was to have a sample size of 100 people, making a total of 300 people. While the structured questionnaire was meant for the mining workers alone whom it was deliberately planned should dominate the sample owing to the size of their population, an interview checklist was used to interview the remaining specialised groups such as service providers, mineral dealers, associations, etc. These people were to be selected by using stratified random sampling involving 6 clusters namely, claim holders, managers (leaders), mining workers, associations, business persons or entities providing services to the mining communities and the neighbouring communities. Stratified random sampling was applied to the mining workers and members of the neighbourhoods depending on the

[9] See Macroeconomic Reforms and Sustainable Development in Southern Africa Project. Field Work Raw Data. Volume 1: Mining Case Study. July-August, 1998.

available blocks and demographic structure of the neighbouring villages. A purposeful sampling procedure was applied to the remaining clusters owing to their small size. Thus this study used a variety of different data collection instruments depending on the cluster.

Nevertheless, some changes were made during the fieldwork. One important change is related to the size of the sample. Following repetitive and/or uniform answers by respondents, it was decided to cut down the initial sample size. Thus, a total of 88 mining workers was interviewed. From the specialised groups, 15 mineral dealers out of whom 6 are also claim holders and 6 others running lapidaries were interviewed. Furthermore, 5 additional claim holders were interviewed. Also in the interview list were 6 regional government officials e.g. Regional Planning Officers and Regional Natural Resource Officers, 8 zonal mines officers, 12 officers of different associations, 25 different service providers and 30 members of the village leadership.

Instruments of Data Collection

This inquiry was based on formal and informal interviews at different levels namely, government departments at all levels i.e. national, regional, district and village levels; Non-Governmental Organisations (NGOs); claim holders, mining companies, mineral dealers, associations and service providers. Apart from the interviews, discussions with different players in the mining sector, together with a study of the analysis of relevant official reports (publications) and data in order to understand the framework and operations of the mining sector, were carried out.

Data were collected using two major instruments namely; structured (systematic) questionnaire and unstructured interview guide to be administered to individuals. Also focused group interviews were conducted to supplement what individual interviews could not achieve, e.g. serving as instruments for eliciting data which are of interest to a group of persons but not necessarily to an individual, while also serving to check the validity of information provided by individuals. The researchers also made use of field observations and video tapes of field activities and physical features of the research areas to supplement data obtained by way of interviews. The video tapes were mainly meant for the outreach (dissemination) programme planned as part of the project.

Research Questions

A number of research questions were addressed in order to gauge:

- The extent to which the local people are involved in decisions to confer new rights of holding;
- The methods and procedures that are in place to ensure that liberalised mining activities are undertaken with sustainable development in mind;
- The way these methods and procedures are being implemented;
- Existence of criteria or mechanisms for ensuring that external costs emanating from mining operations are internalised;
- The kind of compensation extended to the local people who have been displaced from their income-generating resource base;
- The kind of negative social and cultural impacts caused by mining activities and how these can be mitigated; and
- The environmental impacts that are associated with mining activities in the study areas and how they can be mitigated.

Analytical Framework

The conceptual map for earlier studies on economic reforms and sustainable development (see Bagachwa, et al, 1995; Reed, 1996) used a top-down approach. In this approach, macro-economic policy changes such as the structural adjustment programme, are traced from the impact area (household unit level) to the policy level within the existing institutions and sector policies.

In this study, the analysis begins with observations on the environment, social and economic impacts of mining activities as revealed by survey data, working towards identifying proximate and underlying determinants for the observed facts. Proximate determinants are the immediate factors to which participants in the mining activities are responding, e.g. market signals from product, labour, capital and land markets, policy and institutional failures, e.g. inability to enforce regulations, unclear property rights definitions, corruption, lack of standards, weak enforcement agencies, tax regime, etc. It is these signals (in terms of rewards, incentives, penalties, competition) which ultimately encourage/discourage/influence the behaviour of participants in mining activities, e.g. mining prospectors, owners of mineral rights, mining workers, service providers, neighbouring communities, etc. Such behaviour includes choice of where to mine, type of technology to adopt, source of labour, labour organisation, scale of production, marketing channel, disposal of waste, contractual obligations with workers, tax compliance, attitude towards government agencies, etc.

On the other hand, the underlying factors are those significant causal factors which determine or which are the sources of the signals and other proximate factors that come to individuals, firms and households engaged in the mining sector. These may range from local micro and sectoral policies and practices to national macroeconomic policies as well as global-scale external influences. For example, we know very well that although economic policy reform packages result from national macro-economic imbalances, they are invariably implemented with conditions set by international financial institutions. On the other hand, sectoral policies, e.g. mining, land, environment, trade, rural development, infrastructure development, operate within the bounds set by the macroeconomic policy regime and other exogenous factors such as initial socio-political, economic and environmental conditions.

The survey generated both quantitative and qualitative data. The quantitative data were utilised in descriptive statistical analyses using tabulations, cross tabulations, calculation of ratios, percentages, etc., as well as analytical techniques such as comparative and correlation analyses. On the other hand, the qualitative data were analysed to try and discover linkages and structure-function relationships. The analytical framework is shown in Figure 2.1.

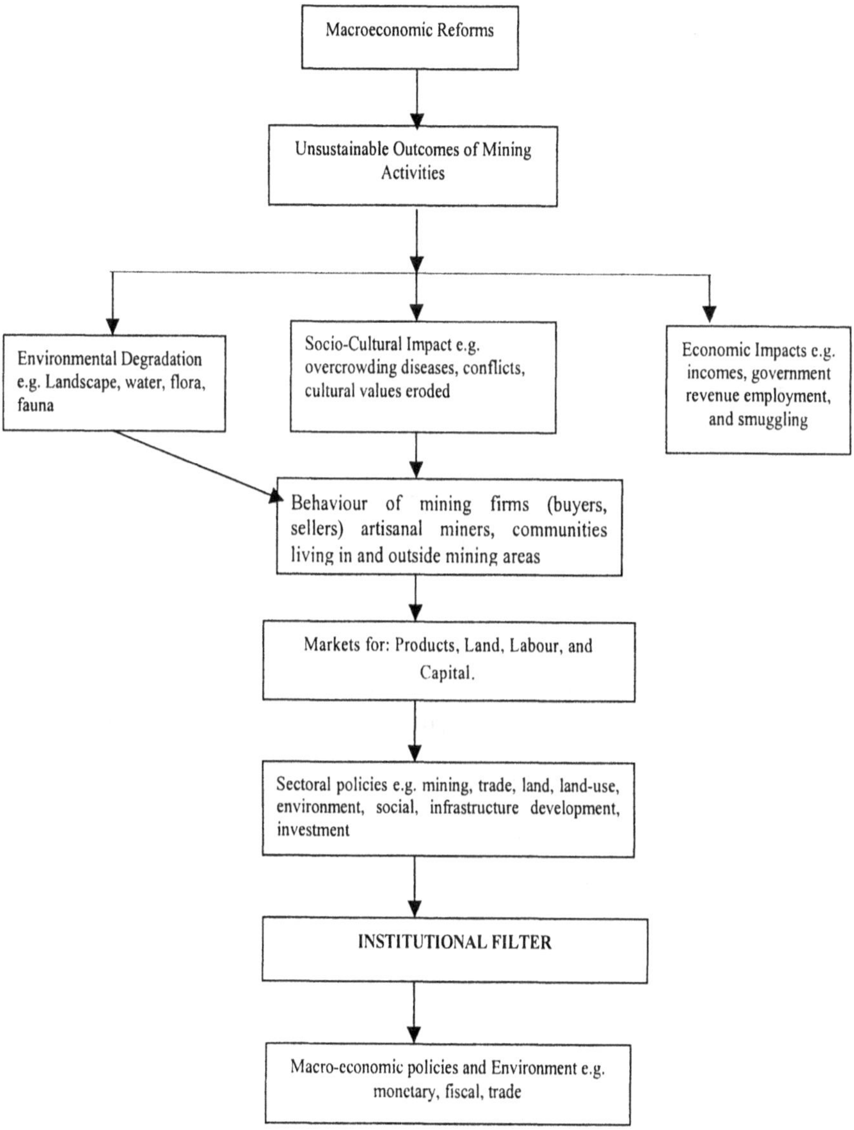

Figure 2.1:*Analytical Framework for UnderStanding the Relationship Between Macro-Economic Reforms and Sustainable Development: The Case of Mining in Tanzania*

3

MINING DEVELOPMENT AND ECONOMIC GROWTH

The overall objective of this study has been to analyse the macroeconomic policy reforms and the policy instruments deployed by the government towards achieving the set objectives. The intention is to assess the subsequent response of the key players in the mineral sector and investigate the impacted areas (gainers and losers). In other words, the inquiry is meant to make an assessment of the expanding investments and operations in the mining sector as it affects the economy, the environment and the welfare of the people both positively and negatively. The primary goal was to establish whether or not the expanding mining sector ensures a sustainable development path in Tanzania. A number of research questions and research hypotheses were developed to guide our investigation. This section presents the findings of the research based on the framework of the three pillars of sustainable development namely, economic, environmental and social.

Labour Supply

The total number of employees under one claim title varies from time to time depending on the conditions of work and productivity of the mining pits. Approximately 20 to 60 employees are employed in one claim title. The number of employees tends to increase during peak production period and shrinks during the low season. Most of the labour comes from the neighbouring districts. Table 3.1 provides figures on sources of labour for the three research areas. Out of the total sampled mining workers at Mererani (Arusha) 41% indicated that they come from other districts within Arusha region. Approximately 27% come from the neighbouring regions, 24% from distant regions and 8% from the same district (Simanjiro).

Table3.1: *Labour Source and Supply in Selected Mining Sites -1998 (in percentage)*

Sn.	Place of Origin	Arusha	Mwanza	Tanga	Row Total
1.	The same locality	2.7	0	0	1.2
2.	The same District	5.4	41.3	31.2	22.4
3.	Other Districts	40.5	23.5	34.4	35.3
4.	Neighbouring Regions	27	17.6	12.5	20.0
5.	Distant Regions	24.4	17.6	15.6	19.8
6.	Outside Tanzania	0	0	6.3	6.3
7.	Total	100	100	100	100

Source: *Survey Data (1998)*

Out of the total sampled employees in Geita (Mwanza) 24% are from neighbouring districts, 41% from Geita district and 18% from the neighbouring regions. The remaining 18% are from distant regions. All of the interviewed employees are males. Nevertheless, a few female employees are also present, mainly in much more lighter jobs such as clerical and catering services.

Approximately 34% of the sampled workers at Umba come from the neighbouring districts, 31% come from Korogwe district where most of Umba Mining area is located and 13% come from the neighbouring regions. About 16% of the sample employees are from the distant regions and 6% are foreigners. The foreigners work with Africa-Asia Precious Stones (AAPS) mining company. Female labour consists of 19% of the sampled employees. Unlike Mererani, in the Umba and Geita mines the participation of women in mining activities is very significant although the sampled population does not show this. In Geita, women are mainly involved in all the steps after hoisting, i.e. in more or less the entire process of gold benefication, while at Umba they participate mainly in screening and sorting. These are lighter jobs compared with excavation and hoisting.

Table 3.2: *Demographic Structure of the Sampled Population in the Research Areas - 1998*

Sn.	*Age (Yrs)*	*Arusha*	*Mwanza*	*Tanga*	*Row Total*
1.	18-25	21.1	0	12.5	13.8
2.	25-30	23.7	29.4	15.6	21.8
3.	30-35	15.8	17.6	25	19.5
4.	35-40	15.8	17.6	25	19.5
5.	40-45	10.6	11.8	6.3	9.2
6.	45-50	2.5	11.8	6.3	5.7
7.	50-55	2.6	5.9	0	2.3
8.	55 or more	7.9	5.9	9.4	8.0
9.	Column Total	43.7	19.5	36.8	10.0
10.	Male	94.7	100	81.3	90.8
11.	Female	5.3	0	18.8	9.2

Source: *Survey Data (1998)*

Owing to the nature of mining work and the general working conditions, almost all the employees are male. Nearly 95% of the sampled employees at Mererani are men (see Table 3.2). Sex and therefore age are said to be the most important demographic factors influencing labour supply in the mining activities. Most young men between 18 and 45 years of age migrate to the mining areas in search of employment as can be seen from Table 3.2. Labour migration towards attractive pastures was common mainly during the colonial period. It normally involved young and productive segments of the population and therefore the overall availability of labour in the source areas is massively diminished which tends to affect agricultural production. However, the impact of such labour movements was not the focus of this investigation.

Individual Income Generation From Auxiliary Activities

As noted, mining creates gainful employment opportunities in the country either directly or through auxiliary activities, which are mushrooming in the mining communities. The major service-providing activities include food and beverages such as restaurants and agents in soft drinks and alcohol, religious institutions, health centres, guest houses, workshops, commodity shops (markets) and transportation.

The second half of the 1980s especially saw the influx of small-scale mining activities throughout the country. Gradually, these mining communities started to expand, creating markets for food as well as for other manufactured products. Agriculture in the neighbourhood of the mining villages benefits from the sale of their products. Traders from small towns located up to 100 kilometers from the mining villages increase their trade volume by taking their merchandise to the make-shift towns in the mining areas. Liberalisation of the mineral sector has thus created gainful employment in auxilliary activities such as trade in agricultural produce, household needs and transport. A few areas which have benefited in this way include Umba in Tanga, Lusu in Nzega and Mererani in Arusha. A significant part of the trade is informal and therefore records are almost non-existent.

In the 1990s Tanzania registered many new investments in the mineral sector most of which are large scale. In Nzega and Geita for example, large foreign companies made an impact on market demand in the mining areas following a change in taste caused by the newcomers. Market prices increased dramatically in these areas following the increased demand and change in consumption pattern and product mix. House rents, transport costs and consumer prices increased by an average of 150% over prices prevailing in the 1980s in most of the mining communities and their integrated markets. While smallholder farmers in the vicinity and traders made good profit margins, the ordinary people were hard hit.

Price signals tended to exert more pressure on natural resources and the environment as charcoal and firewood traders and the construction industry by and large exploited the surrounding forest resources and reserves.

In general, economic liberalisation has led to the expansion of economic activities in the mineral sector. The number of auxilliary activities and trading products has also increased thus creating employment opportunities. The impact of the growing trade volume is also felt in other parts of the country. Industrial merchandise from Zanzibar, for example, finds its way to Mererani in Arusha and Umba in Tanga.

Movement of People into Mining Areas

Respondents were asked to state when they moved to the mining site in order to get an idea of the time period when they were attracted into mining activities. It is noted that there are notable variations as to when most people moved into the three mining areas. In Arusha, 50% of the respondents moved into Mererani area between 1990 and 1995 and about 29% moved slightly earlier, e.g. during the period between 1985 and 1990. Only about 10% moved in between 1980 and 1985. Prior to 1980, only 5% moved into the area and an equal response applies to the period after 1995. Thus, it is noted that 79% of the respondents moved into Mererani in the ten-year period between 1985 and 1995. This is at the height of the economic reform period.

In the case of Mwanza, results show that very few people (5%) moved into the mining areas before 1980, while no-one moved during the period 1980-1985. However, during 1985-1995, the majority of them (70.6%) moved into the gold-mining areas of Geita. It is further observed that 23.5% of the respondents continued to move after 1995, reflecting the upsurge of gold-mining activities in these areas after 1995.

On the other hand, movements into the gemstone-mining areas of Tanga progressed more or less steadily over time. Whereas before 1980 only 6.3% of the respondents moved into the area, between 1980 and 1990, nearly 31% of the respondents did so. Between 1990 and 1995, a further 22% moved in. The biggest influx into the area occurred in the period after 1995, in which 41% of the respondents moved into the gemstone-mining areas of Tanga. This finding suggests that, during the period of economic reform (1985 onwards), a lot more people moved into mining areas to seek better opportunities for earning an income.

Change of Occupation

Most of the interviewed miners were formerly employed in fairly good jobs. Table 3.3 lists occupations of the sampled mining workers before joining the mining sector. Approximately 32.8% of the total sample used to work as farmers and 20% of the sampled miners were civil servants working as teachers, office clerks and technicians. Other previous occupations with the respective percentage of total sample in brackets include mining (18.8%), trading (8.2%), students (4.7%), carpentry (3.5%) and security guards (2.4%), while plumbers, bakers, evangelists, hotel attendants, charcoal dealers and tailors account for 1.2% each. The reasons given for moving into the mining area are also interesting. In Arusha, 50% mentioned joining mining activities as the principal reason for shifting to Mererani, while in Mwanza it is 70.6% and Tanga 84.4%. However, another reason is also mentioned, e.g. looking for a better workplace. Arusha has the highest response (45%), followed by Mwanza (30%) and Tanga (6.3%).

Table 3.3: *Occupational Background of Miners in Mining Areas in Tanzania (1998)*

Sn.	*Occupation*	*Arusha*	*Mwanza*	*Tanga*	*Row Total*
1.	Civil worker	9	2	6	20%
2	Plumber	0	0	1	1.2%
3	Mining	3	7	6	18.8%
4	Farming	11	7	10	32.8%
5	Baker	0	0	1	1.2%
6	Student	3	0	1	4.7%
7	Unemployed	1	0	1	2.4%
8	Evangelist	0	0	1	1.2%
10	Private Traders	6	0	1	8.2%
11	Security Guard	1	0	1	2.4%
12	Charcoal Dealer	0	0	1	1.2%
13	Tailoring	0	1	0	1.2%
14	Carpentry	2	0	1	3.5%
15	Column Total	36	17	32	100%

Source: *Survey Data (1998)*

Employment Contracts

The most dominating mode of payment is production-sharing whereby a certain percentage of the total harvest is paid to workers. The survey data reveal that about 66% of the workers are paid on the basis of sharing the output, 26% earn their income by directly selling the minerals to the brokers, claim holders or mineral dealers and 8% are paid a monthly salary. In Geita, production-sharing payment criteria are very clear as miners organise themselves into groups and mobilise capital, working tools and food before entering either into written or verbal agreement with the claim holder who allows the group to work in a claim holder's plot. Distribution of the output is such that 30% of the harvest goes to the claim holder while 70% is shared among the group members based on their participation and overall patronage. At Mererani and Umba, production-sharing is not clear and the sharing and/or output proportions distributed among the key players vary from one claim holding to another. It was found that only 4% of the employees under production-sharing had a written employment contract, the remaining 96% work under a verbal agreement, which does not guarantee them job security.

The major problems relating to the production-sharing system include production stoppages, exploitation and job insecurity. When there is no production, mining workers have no income implying that long periods of production stoppages can have a severe effect on mining workers which also increases crime. The problem of exploitation is particularly evident in Geita, where claim holders do not directly engage themselves in production. They are actually masters who do nothing apart from waiting for the harvest. Mining workers do all the major work, including mobilisation of the working capital. What a claim holder does is to provide administrative and monitoring services. Furthermore, production sharing does not guarantee job security especially for those who have no written agreement. When workers' employment is threatened, the legal machinery is not in a position to defend them or protect their interests because the workers do not possess any document with legal status.

Claim holders and/or employers do not like payment based on salary schemes because, they argue, it encourages workers to put very little effort into their work, but for workers this scheme is probably better because it guarantees them an income on a monthly basis. Production-sharing on the other hand does not guarantee workers a steady flow of income because a claim holding and/or a mining pit may stop production for a long period during which workers have no income. However, some fortunate mining workers can earn millions of shillings during peak periods of production. The flow of the earnings is therefore highly fluctuating and uncertain. Those who are neither entitled to monthly income nor to production-sharing are reportedly ignored by their employers particularly during harvesting period, thus prompting them to try and divert some of the output from the pit to other destinations, without the knowledge of the owner.

Despite harsh working conditions and many other obstacles in the mining communities, 43% of the interviewed workers reported that they are currently doing fairly well in terms of the income earned and the ability to meet their daily requirements compared with the situation they were in previously. Workers of the mines have repeatedly acknowledged that mining activity is a gamble in that production and the flow of income is highly erratic.

A broad-based procedure of acquiring title deeds in the mineral sector which allows traditional villages to participate in the process of approving the licence can provide a reliable source of income for the local authorities and therefore the indigenous people through levies. This is an important source of income, which, as noted earlier, in some areas such as Tanga Region, has already been formalised, and in some villages, namely Chongoleani and Magamba in the same region, it has started to make an impact.

Creation of Employment Opportunities

Table 3.4 provides estimates of the total number of individuals employed by small-scale miners in the surveyed regions between July and December 1998. Note that these figures do not include employment in auxiliary activities and large-scale mining. The conservative estimates amount to an average of 30 individuals per claim title. Statistics in Table 3.4 depict that only 50% of the potential employment is filled. The under-utilisation of the existing capacity is due to the fact that almost a half of the claim titles available in the three regions are not operating because of capital deficiency.

Table 3.4: *Estimated Total Employment of Small-Scale Miners in Selected Regions in Tanzania: July-Dec. 1998*

Sn	*Regions*	*Total No. of Claim Title*	*Total No. of Operating Claim Titles*	*Average Labour employed*	*Total Employment*	
					Potential	*Actual*
1.	Arusha	600	342	30	18000	10260
2.	Mwanza	550	400	30	16500	12000
3.	Tanga	600	215	30	18000	6450
Grand Total		1750	957	30	52500	28710

Source: *Survey Data (1998)*

Lack of resources to purchase the necessary equipment is one of the leading factors which has forced many investors to suspend their operations. In some areas such as Umba in Tanga region, where alluvial mining is dominant, productivity of the mining plots has been declining and the present activities are taking place in areas that have repeatedly been mined in the past. Since productivity is very low, claim holders and the labour force have tended to abandon such areas. Unless they acquire sophisticated equipment to undertake open cast mining, the possibility of working on those plots again is very slim. This is what has prompted the miners to move from one area to another. In other words, the miners continue working in one area as long as the marginal revenue (MR) is higher than marginal cost (MC). When the MR falls below MC, the tendency has been to abandon the area and search for new ones. This is a typical of Umba, Geita and Matabe. The point of exit arrives early due to the use of crude and inferior technology.

Restructuring of the mining sector has tended to expand mining activities in the country (Nanyaro, 1994; Mutagwaba, *et al.* 1997). As noted before, the number of mining companies, claim holders, auxiliary activities and new discoveries of mineral reserves has increased dramatically. Close to 80% of the mineral revenue used to be contributed by small-scale artisanal miners. This ratio has already changed following the advent of large-scale mining companies.

The mining sites owned by mining companies have created potential markets especially for home-grown agricultural products which can be captured by indigenous producers in the neighbourhood of the mining sites thus, improving their incomes. Unfortunately, very few farmers have realised this, not yet starting to benefit from these opportunities. For vegetables and fruit for example, most mining companies do not rely on the local producers because they are unreliable. Ashanti Gold Mines Company in Geita gets its supplies from Mwanza which is located almost 100 kilometers away, while there is ample agricultural land in the neighbouring villages.

As a labour-intensive activity, small-scale artisanal mining appears to have a greater impact on employment than large-scale mining despite the fact that the rate of employment varies from time to time depending on the prevailing policies.

Lwakatare cited in Mutagwaba *et al.* (1997) estimated artisanal mining to consist of 200,000 or more individuals who are working full time as well as on a part-time basis. Chachage (1995) argues that small-scale miners have made and continue to make new discoveries in many parts of the country every year to the extent that in 1992 they operated nearly 500 gold-mining sites and 300 gemstone sites. By 1993, it was estimated, there were 1,440 small-scale claim holders and 480 prospecting certificate holders in the country. The statistics by the Federation of Miners Associations of Tanzania (FEMATA 1997) indicate that there were no less than 550,000 individuals involved in mining of minerals as small-scale artisanal miners. Current FEMATA (1999) estimates indicate a total of more than 1,000,000 individuals engaged in mining.

With the exception of Chachage (ibid.) and FEMATA (ibid) who consider both individuals employed in small-scale mining and auxiliary activities, others appear to consider employment in small-scale artisanal mining activities alone, i.e. excluding service providers. However, judging from estimates by Lwakatare (1993), Tan-Discovery (1996), FEMATA (1997) and FEMATA (1999), it can be concluded that, during the reform period, employment in the mineral sector expanded. Nevertheless, the methods used in the reviewed studies are based on estimations. Many senior executives of FEMATA and TAMIDA have cautioned that these estimates are likely to understate the actual situation.

Tax Regime for the Mineral Sector

The contribution of the mineral sector to the growth of the national economy has not only been through employment generation. Other contributions are evident through tax revenues, licence fees and renewal fees paid by investors.

Tax Structure Prior to 1997/98 Financial Year

Prior to 1997/98 the list of government taxes for the mineral sector was too long and changed frequently. The structure that prevailed before 1997/98 included the following taxes:

Stamp Duty:	1.5% of the value of minerals purchased
Withholding tax:	2% of the value of minerals purchased
Royalties:	3% of the export value of rough stones and
	1% of the export value of cut stones
Sales tax:	3% of total sales
Corporate tax:	35% and 40% (for residents and non-residents respectively) value of gemstones and 5% of the diamond export value

This regime was characterised by tax evasion thus denying the government much needed revenue. Most mineral dealers did not have the incentive to operate officially and thus illegal market channels became an attractive alternative for disposing of mineral outputs. The country is believed to have lost huge amounts of revenue through illegal trade and tax evasion during this period.[10]

This period was also characterised by rather tedious and bureaucratic procedures to process master dealer licences. Applicants were supposed to deposit TShs. 5 million in a bank before processing their applications.

[10] This is based on the interview with TAMIDA (1998) and FEMATA (1999).

In every region, regional committees had to convene to discuss and approve the applications. The meeting was normally held after the regional mining officer had received the application and indicated to support it. The regional trade officer used to chair the meeting while the regional mining officer was the secretary. Other members of the meeting included the regional security officer, regional crime investigation officer, regional co-operative officer and the bank manager. Costs of the meeting were to be met by the applicants. Thereafter, the application went through the Ministry and finally to the State House (President's Office) for final approval. This bureaucratic procedure created room for corruption, thus crowding out many possible applicants.

Tax Structure Since 1997/98

A number of weaknesses and pitfalls in the previous system were addressed in the 1997/98 budget proposal of the Ministry of Minerals and Energy. These included, abolition of some of the taxes, the bureaucratic regional committees and the TShs 5 million deposit (URT, 1997b). Today master dealers are obliged to pay the following taxes only:

- Royalties: 3% of export value of rough stones (to encourage domestic processing industries, i.e. Lapidaries);
- Stamp Duty: 1.2% of the value of purchased minerals; and
- Withholding tax: 2% of the value of purchased minerals.

Dealers have described the present tax structure as being far less arduous compared with the previous one. Following the recent policy and legislative measures, the process of obtaining mineral rights has been significantly simplified. Furthermore, under the 1973 Income Tax Act and the National Investment (Promotion and Protection) Act 1990 (NIPPA), administered by Tanzania Investment Centre (TIC), both local and foreign investors are entitled to a tax holiday and a few tax exemptions.[11] The incentive package is obviously very attractive. Subsequently, more applications for mineral rights have been processed and the number of new investors has increased significantly.

---

[11] See also Davis et.al. (1997) and the amendment made during the 1999/2000 budget. However, as we shall see later, some articles of the Act have been amended in 2002.

Table 1.2 shows the extent to which the number of licences increased between 1990 and 2002. Table 3.5 shows the number of investments approved by Tanzania Investment Centre (TIC) between 1990 and 2000 With the exception of 1995, in each year there has been new investment in the country which generates employment and revenue through licence fees, rental fees as well as renewal fees. Various rents, fees and other charges collected by the government are presented in Table 3.6.

Statistics on tax revenue collected from the mineral sector are disorganised and/or absent for most transactions. Nevertheless, perhaps the most important at this stage is to acknowledge that the growth of formal investments in the sector has the potential of realising increased tax revenue from mineral trade as can be depicted from Table 3.8 below. This table shows clearly that royalties have grown by more than 800% between 1996 and 2000!

Table 3.5: *Approved Projections, Employment and Total Value of Investment in Mining in Tanzania (1990 – 2000)*

Year	*Investments*							
	Total Projects	New Projects	Old Projects (Expansion & Rehabilitation)	Domestic Projects	Foreign Projects	Joint Venture	Employ-ment	Value of Investment (Million Tshs)
1995	0	0	0	0	0	0	0	0
1996	1	1	0	0	0	1	30	4,680
1997	14	14	0	8	2	4	2,248	173,353
1998	8	5	3	4	2	2	2,283	25,128
1999	17	14	3	5	6	6	1,022	27,560
2000	76	65	11	27	14	35	9,076	417,600
1990-2000	128	111	17	44	24	60	16,134	656,015

* These are investments approved by Tanzania Investment Centre and include Petroleum Products and Minerals

Source: *Tanzania Investment Centre (TIC) and URT: Economic Survey (1999)*

Table 3.6: *Preparation and Annual Rental Fees on Mineral Rights (Licences) in Tanzania (1998)*

Sn.	*Type of Licence*	*Local Company (Tshs)*	*Foreign Company (US $)*
A.	**Annual Rental Fees per square kilometers**		
1.	Reconnaissance Licence (RL)	500	10
2.	Prospecting Licence (PL)	2,000	30
3.	Mining Licence (ML)	150,000	1,500
B.	**Preparation Fees**		
1.	Reconnaissance Licence (RL)	20,000	250
2.	Prospecting Licence (PL)	30,000	400
3.	Mining Licence (ML)	50,000	600
C.	**Renewal Fees**		
1.	Reconnaissance Licence (RL)	10,000	200
2.	Prospecting Licence (PL)	10,000	200
3.	Mining Licence (ML)	10,000	200
D.	**Fees for transfer Approval**	10,000	200
E	**Fees for certificate of surrender**	10,000	200
F	**Mineral dealer Licence (MDL)**		
1.	Master gemstone Dealer	200,000	-
2.	Gemstone Lapidary Dealer	120,000	-
3.	Master Gold Dealer	200,000	-
4.	Master Diamond Dealer	200,000	-
5.	Licence to Purchase Industrial minerals	60,000	-
6.	Assayer's licence (Gold and Silver)	200,000	-

Source: Kayera *et. al.* (1996) and TIC (1998)

Table 3.7: *Advanced Mining Projects in Tanzania*

S/No	*Company*	*Area*	*District*	*Mineral*	*Estimated Resources and/or Reserves (Ounces)*	*Production Per Annum (Ounces)*	*Investment Capital (Mill. US$)*	*Production Commencement Dates*
1.	Resolute Ltd.	Lusu (Golden Pride)	Nzega	Gold	2,800,000	220,000	72	November 1998
2.	Kahama Mining Corporation Ashanti Goldfields Ltd.	Bulyanhulu	Kahama	Gold	10,400,000	400,000	350	First quarter of 2001
3.	Anglo Gold	Nyankanga	Geita	Gold	12,100,000	500,000	165	June 2000
4.	Afrika Mashariki Gold Mines	Nyambirama & Nyambigena	Tarime	Gold	4,100,000	200,000	90	Mid 2001
5.	Kahama Mining Corporation Pangea Minerals	Nyalingo (Golden Ridge)	Shinyanga	Gold	2,200,000	60,000	37	2002
6.	Madaba/Anglo-American/ Pangea Minerals	Mwime (Chocolate Reef)	Kahama	Gold	2,000,000	-	-	2002
7.	East Afrika Mines	Buckreef, Rwamagaza, Buziba, Nyakafuru	Geita	Gold	375,000	-	-	Investigation in progress
8.	Pangea Minerals Ltd.	Mgusu	Geita	Gold	271,000	-	-	Investigation in progress
9.	Pangea Minerals Ltd.	Tulawaka	Geita	Gold				Investigation in progress
10.	Anglo-American/El-Hillal	Nyamulilima	Geita	Gold	2,250,000	-	-	Investigation in progress

S/No	*Company*	*Area*	*District*	*Mineral*	*Estimated Resources and/or Reserves (Ounces)*	*Production Per Annum (Ounces)*	*Investment Capital (Mill. US$)*	*Production Commencement Dates*
11.	Merelani Mining Ltd.	Merelani	Simanjiro	Tanzanite, Graphite				Mine development under way
12	Williams Diamonds Ltd. (DeBeers $ Tanzania Govern)	Mwadui	Shinyanga	Diamonds		226,780 carats		Producing
13	Kiwira Coal Mine Ltd.	Kiwira	Kyela	Coal	35,500,000	50,000 tonnes		1998
14	National Development Corp.	Mchuchuma-Katewaka	Ludewa	Coal Iron ore	535,000,000			Joint Venture Partner Identified
15	National Development Corp.	Liganga	Ludewa	Vanadium & Titanium	200-1200Mt (45Mt proved by drilling) Fe203-51.4%, FeO-16.9% TiO2-13.0, V205 0.6-1.7% A1203-8%, MgO-4.8% SiO2-0.9%			To conduct the following: Study for establishing flow sheet, cost estimates for supplementary exploration and reserve estimation Preliminary capital and operating costs Market demand and price trend study

S/No	*Company*	*Area*	*District*	*Mineral*	*Estimated Resources and/or Reserves (Ounces)*	*Production Per Annum (Ounces)*	*Investment Capital (Mill. US$)*	*Production Commencement Dates*
16	Kabanga Nickel Co.	Kabanga	Ngara	Nickel	21.27Mt at 2.18% Ni, 0.27% Cu & 0.16 Co.	12,000 (a)	-	2001
17	Gammon Tanzania Ltd.	Madaba	Songea	Uranium				Investigation in progress
18	Tanzania Portland Cement	Wazo Hill	Kinondoni	Limestone Clay				Producing
19	Mbeya Cement	Songwe	Mbeya	Limestone Clay				Producing
20	Tanga Cement	Tanga	Tanga	Limestone Clay				Producing

Key
a) - tonnes concrete
b) - 1 million troy ounces of gold equivalent to 31.1 tonnes

Source: *Ministry of Energy and Minerals*

Table 3.8: *Government Revenue Accruing From Royalties (1996 – 1999)*

Sn	*Category*	*1996 (Millions)*		*1997 (Millions)*		*1998 (Millions)*	
		TShs	US $	TShs	US $	TShs	US $
1.	Gemstone	117.7	0.203	89.7	0.147	82.3	0.136
2.	Diamond (Export by WDL)	n.a.	n.a.	n.a.	n.a.	313.5	0.517
3.	Diamond (Export by Dealers)	14	0.024	20.7	0.034	21.2	0.035
4.	Gold (Export by Dealers)	n.a.	n.a.	11.7	0.019	3.7	0.003
5.	Gold (Export by Large-Scale Miners)	n.a.	n.a.	n.a.	n.a.	0	0
6.	Gold (Export by Meremeta Ltd)	n.a.	n.a.	n.a.	n.a.	n.a.	n.a.
	Total	131.7	0.227	122.1	0.199	420.7	0.691

Sn	*Category*	*1999 (Millions)*		*2000 (Millions)*	
		TShs	US $	TShs	US $
1.	Gemstone	471.8	0.590	471.8	0.590
2.	Diamond (Export by WDL)	674.9	0.844	1,115.7	1.4
3.	Diamond (Export by Dealers)	22.8	0.029	53.1	0.66
4.	Gold (Export by Dealers)	15.6	0.02	10.0	0.013
5.	Gold (Export by Large-Scale Miners)	942.3	1.178	4,813.0	6.0
6.	Gold (Export by Meremeta Ltd)	n.a.	n.a.	n.a.	n.a.
	Total	1,857.7	2.324	6,463.6	8.663

Notes: Large scale miners comprise of Golden Pride Project and Ashanti Gold Fields - Geita. Also note that, following the difficulties involved in obtaining revenue data from Meremeta Limited, royalties paid by this company is not available despite the fact that the company makes significant gold exports.

Source: *Ministry of Energy and Minerals*

Note the clear link between the reform programme to liberalise the economy and the growth of economic activities in terms of the number of investments in the mineral sector. This link confirms the hypothesis put forward earlier on the positive relationship between the two variables.

Tax Revenue From the Mineral Sector

Tax revenue collected by the government from the mineral sector has been difficult to characterise following problems of capturing the mining rents due to marketing problems. Revenue contribution is likely to increase dramatically, particularly from the year 2000, following the opening up of many gold-mining companies whose total annual production will be not less than 20 tonnes. These companies are listed in Table 3.8.

Since commencement of its production in November 1998 up to March 2000 Resolute Limited has produced a total of 6.63 tonnes (213,183 ounces) of gold valued at US $ 55.65 million. The government had a deliberate policy of not fully involving itself in administering tax in the so called "informal mining sector.[12]" There are various reasons but the most conspicuous one is the allegation that small-scale miners and/or artisanal miners are very sensitive to government policies and difficult to handle. They are many, scattered almost all over the mining regions and most of them are unregistered. Tax policies targeting such groups will therefore be very inefficient and expensive. One of the executives wrapped it up by noting that:

> "So far mining has not been an important sector to the nation and particularly in terms of its revenue contribution because most of the production and trade is informal. Tax administration is extremely costly. Unless the "informal mining" is transformed, we are not planning to deploy expensive resources in order to fully capture whatever this sector is generating, at least not for the time being when "informal small-scale miners" dominate production. In future

[12] Some stakeholders, particularly many of the senior and top government executives, refer to small-scale miners as informal miners.

> we are determined to get in because large-scale miners who are much more organised and easy to administer will be contributing over and above what is contributed by the informal mining sector" (1999).

As one of the policy measures to correct this discrepancy, the government has made investment atmosphere for large-scale miners more attractive. The hypothesis is that, unlike informal miners, large-scale miners are formal, efficient, productive and manageable. This concurs with the campaign launched by the government to convince small-scale miners either to surrender their claim titles or to enter into a joint venture with large-scale miners. These campaigns were evident in Mgusu and Rwamgasa mining villages in Geita districts.

Thus the government's plan is to increase or improve and direct its tax collection efforts towards large-scale miners some of whom have just started production as shown in Table 3.8. Due to the government's reforms, tax revenue from this sector has improved tremendously. Table 3.7 shows government revenue from royalties between 1996 and 2000.

The government revenue generated from royalties declined between 1997 and 1998 partly because of the poor tax structure instituted by the government prior to the 1997/98 changes. After the 1997/98 tax reforms and the commencement of production by some mining companies, royalties grew from Tshs 122.1 Million in 1997 to TShs 6,463.6 Million in 2000.

Illegal Cross Border Trade (ICBT)

In Tanzania, the official statistics suffer from two major weaknesses, namely, incompleteness as well as inaccuracy (Ackello-Ogutu *et al.*, 1998). It should be borne in mind that one of the causes of this discrepancy is the unrecorded trade between Tanzania and the rest of the world.[13]

Although there is a substantial reduction of tariffs and a relatively attractive

[13] According to Ackello-Ogutu (1998) unrecorded trade is broadly defined as all trade activities which should be included in the national income but are not captured by official national statistics. The omission of such activities could easily lead to faulty policy analysis.

business environment, Tanzania's minerals are still siphoned off. Smuggling and cheating remain widespread. Some outsiders regard Tanzania as a "*terra nulli*" (nobody's land) because the government is said to be overly liberal and generous to foreigners.

In northern Tanzania, minerals are smuggled to the neighbouring countries of Kenya and Uganda. Some of the minerals find their way to Burundi, Rwanda and the Democratic Republic of Congo. In the south, Zambia, South Africa and Malawi are mentioned as among the attractive markets for Tanzanian minerals. Other markets include the Asian continent and Europe through Dar es Salaam International Airport where foreign gemstone dealers especially, prefer leaving the country on weekends, when security checks are believed to be rather loose (Sunday Observer, 1998).

Measures taken by the government to curb smuggling so far include a thorough and comprehensive study on the mineral sector tax differential with neighbouring countries which will be conducted by a local consultancy firm. In addition to that, the government through the Tanzania Revenue Authority (TRA), the Ministry of Finance, the Planning Commission and the Ministry of Energy and Minerals was examining the existing tax rates, with the aim of stamping out the smuggling of minerals. Differences in tax rates in the mineral sector between Tanzania and neighbouring countries largely account for the smuggling of valuable minerals to these countries.[14]

Apart from the overly liberal and generous environment, the brokers, most of whom are unregistered, have to a large extent encouraged illegal exports. Brokers are prohibited from the mining pits by law but in practice they visit these areas in large numbers. This is particularly true in Mererani, Rwamgasa and to a greater extent in Kalalani and Mugusu. Many of the brokers are foreigners who are crossing the border freely from neighbouring countries. Kalalani and Kigwasi, for example, is less than five hours' cycling from the Kenyan border. Village leaders observed that it is extremely difficult to control the movement of foreigners in the area due to the proximity of the villages to the border and the fact that some of the families extend beyond the Tanzanian border. In fact Tanzanian traders based in Kalalani and Kigwasi villages find it cheaper to purchase supplies from Kenya than

[14] See also the Financial Times (1999a) Issue no 146, Wednesday (11-17), pp.3

Tanga in Tanzania. This complication is also evident in Mererani where brokers of Kenyan nationality can afford to operate under the cover of indigenous tribes, which are also present in Kenya (e.g. Maasai and Luo).

Furthermore, the government does not seem to have the capacity to monitor and control the expanding trade in the mineral sector. The regional mining offices are poorly equipped with very limited resources to undertake effective surveillance of the mining sector. This has aggrevated both the degradation of the environment and loss of revenue through tax evasion and illegal cross-border trade.

Table 3.9 gives figures on commercial banks' lending to the mining and manufacturing sectors as well as central government budgetary allocation to the mining sector. The figures are expressed in real terms using the National Consumer Price Index (NCPI: December 1994=100). As can

Table 3.9: *Trends in Government Spending and Commercial Banks Lending to Mining and Manufacturing Sectors in Tanzania: 1992-2001 (Tshs. Mill.)*

Year	*Commercial Banks' Domestic Lending (Mining and Manufacturing)*	*Central Government Expenditure (Mining)*
1992	757	
1993	750	
1994	806	
1995	478	5.5
1996	266	11.8
1997	194	4.2
1998	287	15.2
1999	410	11.2
2000	441	35.7
2001	519	39.4

Source: *URT (1999) Economic Survey 1998.*

be seen from Table 3.9, the real value of financial resources allocated to the mining sector has generally been declining over time despite the fact that the number of approved licences and new investments as indicated in Tables 1.2 and 3.5 has increased, suggesting that the regional mining offices need more resources to be able to enforce governing rules and regulations.

Poor institutional co-ordination is another reason for the negative effects on the sector. During the 1997/98 budget session, some of the taxes such as stamp duty and royalties were adjusted downwards and the mineral dealers perceived it that way. However, since the Finance Bill had not been amended accordingly, the Tanzania Revenue Authority (TRA) has been pressing mineral dealers to continue paying the old rates despite these changes.

The controversial tax holidays and exemptions under the 1973 Income Tax Act and the National Investment (Promotion and Protection) Act 1990, administered by the then Investment Promotion Centre (IPC) (now renamed the Tanzania Investment Centre (TIC)), are unnecessarily generous to investors in addition to embodying serious flaws. Guided by the Investment Promotion Committee, the TIC is mandated to award investment certificates of approval to new enterprises, the right to import machines and equipment for the investment at a nominal tax rate of 10% and for the first five years (operative period) free income tax on profits, free withholding taxes on dividends, royalties and interest paid abroad (Davis *et al* 1997). In the case of an investor expanding the existing investment by injecting more capital, the investor gets similar relief from import duty and sales tax, as well as from withholding taxes. The Income Tax Act, 1973 provides for a number of tax relief and investment incentives, including a 20% deduction for the cost of investment in priority sectors such as agriculture, mining, manufacturing, tourist hotels and ships, among other provisions.

There are also a number of tax incentives given under the National Investment Promotion and Protection Act, 1990 including, tax holiday, remission of withholding taxes on interest payments, dividends, royalties, etc. Under the tax holiday, new enterprises (investors) during the implementation period receive the right to import machinery and equipment needed for the investment originally free of import duty and sales tax and since 1996 subject to a nominal tax rate. Furthermore, the investor does not pay income tax on profits and withholding taxes on dividends, royalties and interest paid abroad for the first five years of the operative period (Davis *et al*, 1997).

Apart from discriminating against old investments, the US $ 50,000 capital requirement for domestic investors also discriminates against local investors most of whom have small capital capacity.[15] The response of local investors has consequently been lagging behind foreign investments as demonstrated by the status of reconnaissance in Figure 3.1 below and the appendices 6 and 7. Tax holidays have also tended to encourage concentration of capital where investors are encouraged to speed up and therefore shorten production periods. During the 2000/2001 budget swession, the government made a few amendments aimed at discouraging tax evasion e.g. through changing the names of establishments after the end of every five years hliday so as to qualify for a new (five years) hliday, even when the firm continues with the same fucntions.

	1992	*1993*	*1994*	*1995*	*1996*	*1997*	*1998*	*1999*
Local	5	4	17	3	17	5	19	10
Foreign	8	5	6	2	33	16	47	13
Joint venture	0	0	1	0	0	0	0	0

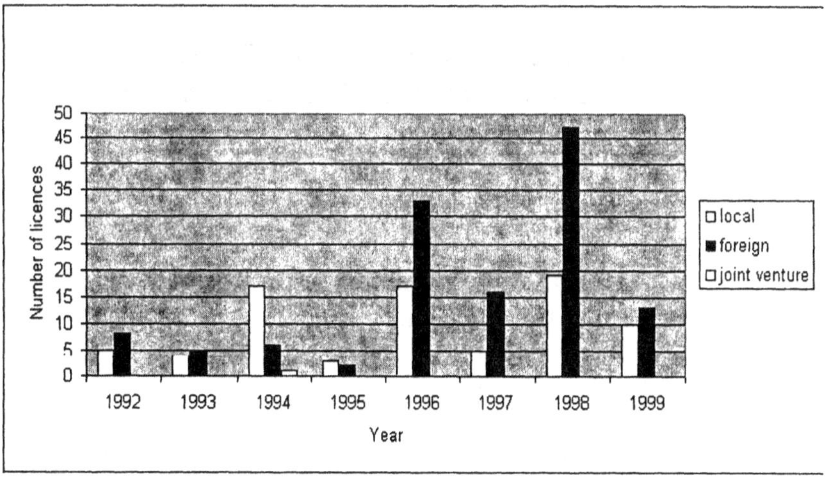

Fig. 3.1: *Status of Reconnaissance Licences Granted Since 1992 up to 30th June 1999*

[15] Initially, the capital requirement for domestic investors used to be US $ 100,000. It has now been reduced though still on the high side.

Economic theory emphasises that, with low tariff rates, there is the likelihood of more compliance, while high duties induce dishonesty and low buoyancy, i.e. more tax evasion (Mpango 1996).

Another important factor relates to the absence of a well-defined market for minerals. There has been an outcry, particularly on the part of the mining workers and small-scale miners, following difficulties in marketing their products. In many cases there are no registered dealers in the mining communities. For example, at Umba one has to travel not less than 73 kilometers from Ng'ombeni to Tanga town where registered dealers are operating, which is costly and risky.

Estimating the Volume of Smuggled Minerals

Smuggling is an illegal exchange of goods undertaken by various mineral dealers, normally taking place via unofficial markets and therefore respective transactions are not recorded. The purpose of this section is to provide estimates to show the extent and direction of the illegal trade in minerals.

Estimates by Chachage (1995) indicate that Tanzania has been producing between 10 and 16 tonnes of gold annually, while the Central Bank and other designated dealers used to purchase an average of 3 to 4 tonnes annually indicating that nearly 10 tonnes of gold has been smuggled out every year. This is equivalent to 73% of the total gold production.[16]

Likewise, 80% of gemstones is believed to be exported informally according to the estimates by Chachage (ibid). Estimates by Nanyaro (1994) show that the proportion of smuggled gold, diamond and gemstones in total production was around 50%.

In 1998 Tanzania exported gemstones worth US $ 8.1 million (TShs 5.6 billion), up from a mere US $ 0.16 million in 1988 according to TAMIDA (1999). Note that, during the state monopoly era when all

[16] This is based on the annual average production of 13 tonnes ((16 + 10)/2) - ((3 + 4)/2) = 9.5, which is 73% of 13 tonnes.

minerals were marketed through a single enterprise, the country hardly managed to export gemstones worth US $ 200,000 per annum. The rest were smuggled out of the country and hence buyers could obtain Tanzania's minerals from a neighbouring country. These figures indicate that the value of smuggled gemstones was approximately US $ 7.9 million.

It was pointed out earlier that the 1979 Mining Act framework was liberalised further during 1987/89 when a gold and gemstone trade rationalisation policy, which permited privatization of mineral trading activities was introduced. Associated with this important change, the Bank of Tanzania and other designated banks entered the gold-buying market, and operated for almost 5 years. During this period the bank used to buy gold by offering world market prices. Thus a significant part of the domestic market was attracted by the world market prices offered by the bank.

Table 3.10: *Tanzania: Gold Production and Value (1986 – 1998)*

Year	*Reported Gold Production (kilogrammes)*	*Value of the Reported Production of Gold (US $)*
1986	47	260,000
1987	201	1,790,000
1988	41	270,000
1989	112	1,152,057
1990	1,643	13,635,060
1991	3,779	29,099,700
1992	4,525	40,380,000
1993	3,370	31,450,000
1994	2,861	27,390,000
1995	320	3,290,000
1996	318	2,780,000
1997	323	1,997,466
1998	427	3,055,379
1999	4,767	39,800,000
2000	15,060	120,570,000

Source: *URT, Economic Survey (Various Issues)*

On the strict assumption that production has been constant, estimated annual average gold production during the period 1986-02 is 3,236 kilogrammes.[17] However, reported and/or official annual gold production figures are given as 100 kilogrammes (1986-89), 3,236 kilogrammes (1990-94) and 347 kilogrammes (1995-98). Note that 1986-89 is the period just before the banks started buying gold, according to Meremeta and Tanzania Chamber of Mines.

Table 3.11: *Estimated Value of Smuggled Gold from Tanzania*

Period	*Estimated Average Gold Production per Annum (kilogrammes)*	*Reported Average Gold Production per Annum (kilogrammes)*	*Estimated Volume of Smuggled Gold per Annum (kilogrammes)*	*Annual Average value of reported gold (in mill US$)*	*Estimated Value of Smuggled Gold (in mill US $)*
1986-89	3236	100	3136	1.48	46.42
1990-94	3236	3236	Minimized	41.39	Minimized
1995-98	3236	347	2889	4.43	36.9

Source: *Computed from Table 3.10.*

As can be seen from Table 3.11, before the banking institutions started buying gold at world market prices, Tanzania was losing approximately 3,136 kilogrammes of gold, which is equivalent to US $ 46.42 million (at current prices) annually due to smuggling. When the bank joined the buying market the volume of smuggled gold was significantly reduced. Afterwards, the volume of smuggled gold started to go up again to an average of 2,889 kilogramms annually, which can be translated into US $ 36.9 million (at current prices). This was the period when the banks had withdrawn from buying gold, i.e. 1995-1998.

One of the driving forces, which has tended to attract illegal mineral exports, is the difference between the domestic price and the world

[17] The figure is derived using production figures for the period 1990-94 when the banks were buying gold. The assumption is that most of the production was reported through the bank.

market price which has always been far above the price offered by unscrupulous dealers and brokers, thus creating an attractive margin which mineral dealers cannot afford to lose. As noted earlier, smuggling of gold and other minerals has been going on even when the banks joined the buying market, although it was relatively reduced. Besides avoidance of tax liability, incentives for smuggling during this period were created by the discrepancies that have continued to prevail between current world market prices and prices offered by the Bank of Tanzania (Chachage, ibid). For example the Bank of Tanzania adjusted its buying prices sluggishly and lagged behind between March 1992 and February 1994.

It also needs to be noted that, when the banking institutions decided to purchase gold and gemstones, they appeared to be performing very well judging from the volume of purchases they were recording. This achievement has prompted questions as to why the Bank of Tanzania (BOT) and other designated banks suddenly opted out of this important and profitable business. The fact is that mineral dealers in collusion with dishonest bank employees got to the habit of cheating the banks by selling fake and contaminated gold and gemstones. It is reported that, the banks used to buy gold consisting of more than 30% impurities and it should be noted that anyone buying such gold will not operate profitably according to the experts working within *Meremeta* which is a gold-buying company.

The preceding analysis shows clearly how the marketing problem contributes to illegal export of minerals from the country.

Production, Export Performance and Marketing

Production and Export Performance

Another area where the sectoral contribution to the national economy can be gauged is revenue accruing from both domestic and international sales.

Due to the expansion of the small-scale miners and liberalisation of the mineral sector, the country's mineral production as well as export volume has been increasing. Diamond and gold continued to be the leading minerals in the country in 1998, 1999, 2000 and 2001, contributing 63.9%, 83.3%, 89.7% and 94% respectively of total mineral exports. Although, gemstones export value have been increasing between 1998 and 2001, its contribution to the total mineral exports have declined from 36% in 1998 to 6% in year 2001.

Williamson Diamonds Limited at Mwadui in Shinyanga Region, which is owned, jointly by the Government of Tanzania and De Beers of South Africa has continued to be the leading producer and therefore exporter of diamonds in the country. Exports increased to 354,388 carats in 2000 compared to 234, 800 carats produced in 1999, but declined to 254,271 carats in 2001. Likewise, gold exports increased from 4767 kilogrammes in 1999 to 14647 kilogrammes and 29520 kilogrammes in 2000 and 2001 respectively.

In 1998 the El-Nino rains heavily disrupted production of gemstones. Subsequently, only 48,518 kilogrammes were purchased compared with 59,590 kilogrammes produced in 1997. Gemstones generated US $ 8.1 million in 1998, well above the US $ 7.9 million in 1997. In 1999 gemstone production increased to 95,200 kilogrammes (equivalent to US $ 14.1 mill), before going down and increasing again to 67,870 kilogrammes (equivalent to US $ 18.5 mill) and 96,266 kilogrammes (equivalent to US $ 18.8 mill).

During the 1980s (1980-89), the contribution of mining to GDP depicted in Table 3.12 has averaged 0.8%, increasing to 1% during the period 1990-92, 1.3 during 1993-98 and 1.5 during 1999-2001. This upward trend signifies the growth of the mineral sector during liberalisation. The last two years, especially, recorded higher rates of mining contribution to the national output at current prices, jumping from 1.5% in 2000 to 1.6% in 2001.

Table 3.12: *Mineral Production as a % of GDP in Tanzania (Mill T.Shs at Current Price)*

Year	*Total GDP*	*Mining and Quarrying*	*% of Mining and Quarrying to GDP*
1965	5,671	139	2.5
1970	8,215	105	1.3
1975	16,988	101	0.6
1980	36,296	209	0.6
1985	166,304	994	0.6
1986	224,464	1,227	0.5
1987	302,680	3,524	1.2
1988	468,100	3,956	0.8
1989	582,166	5,103	0.9
1990	760,005	6,525	0.9
1991	989,594	8,840	0.9
1992	1,275,917	13,503	1.1
1993	1,607,763	19,062	1.2
1994	2,125,324	26,170	1.2
1995	2,796,642	35,190	1.3
1996	3,452,558	38,511	1.1
1997	4,281,600	56,810	1.3
1998	5,125,311	74,386	1.5
1999	5,977,700	85,792	1.4
2000	6,705,134	99,519	1.5
2001	7,590,765	119,422	1.6

Source: TET Vol. 1 No. 4 (1989); TET Vol. 4 No 1 (1991) and URT (1999): Economic Survey (Various Issues)

Likewise, the share of mineral exports contributing to GDP and share of mineral exports contributing to total exports as depicted in Table 3.13 shows an increasing trend. In relation to GDP, mineral exports recorded 3.7% in 1967, 1.5% in 1970 and an average of 0.4% during 1980-89. Thereafter they rose to 1% in 1993 before declining slightly to 0.9 in 1996. During the last two years mineral exports constituted 2.2% and 3.6% of total GDP. The same trend can be gauged in relation to total exports.

Again, the period 1997-98 recorded the highest rates, increasing from 12.9% in 1997 to 15.2% in 1998. Then made a drastic decline to 3.8% and 8.1% in 1999 and 2000 respectively, before rising again to 12.7% in 2001.

The dramatic increase during the last two years partly reflects the contribution of Ashanti Gold Fields, which is jointly working with Resolute Ltd. at Lusu (Golden Ridge) in Nzega district. The company started gold production in November 1998 at an estimated rate of 15,000 ounces each month.

Table 3.13: *Mineral Exports as a % of Total GDP and Total Exports in Tanzania (Million US $)*

Year	*Total GDP*	*Total Exports*	*Mineral Exports*	*% of Mineral Exports to Total GDP*	*% of Minerals Exports to Total Exports*
1967	941.3	244.0	35.2	3.7	14.4
1970	1141.0	246.0	16.7	1.5	6.8
1980	4426.3	508.0	38.9	0.9	7.7
1985	9519.4	324.7	19.6	0.2	6.0
1986	6864.3	346.4	21.5	0.3	6.2
1987	4710.2	346.8	13.4	0.3	3.9
1988	4714.5	372.0	12.8	0.3	3.4
1989	4060.3	414.7	19.8	0.5	4.8
1990	3896.3	402.0	27.0	0.7	6.7
1991	4515.4	361.1	44.0	1.0	12.2
1992	4285.8	397.0	40.8	1.0	10.3
1993	3967.1	367.2	38.8	1.0	10.5
1994	4170.3	519.3	30.0	0.7	5.8
1995	4865.8	661.2	44.9	0.9	6.8
1996	5952.9	768.0	55.9	0.9	7.3
1997	6994.7	717.1	92.8	1.3	12.9
1998	7687.0	676.2	103.0	1.3	15.2
1999	7472.1	2272.0	85.6	1.2	3.8
2000	8360.5	2255.3	181.9	2.2	8.1
2001	8705.0	2459.1	311.9	3.6	12.7

Source: Ministry of Energy and Minerals, Planning Commission, Bank of Tanzania & Tanzania: Statistical Appendix, IMF Staff Country Report No. 98/5, Jan 1998.

Composition of Mineral Production

For a long period small-scale minerals have been dominating the mineral sector in terms of mineral production. A list of large-scale mining companies existing in Tanzania in the past is provided in Appendix 2 and 3. They include Williamson Diamonds Ltd (1940 - to date), Dar Tardine Al-Umma (1984-1989), Placer Dome (1989-1992) and Romanex International (1989-1996). Following the measures taken by the government, which include liberalisation of the sector and putting in place investment incentives, Tanzania has to a large extent succeeded in attracting new large-scale internationally- renowned companies such as Ashanti Goldfields and Anglo-American mining companies. As noted earlier, the gradual participation of large-scale miners will tend to increase and their contribution to the national economy in terms of tax revenue and GDP share will overshadow small-scale miners' contribution.

Unofficial Markets

As with tax revenue, official statistics on mineral production as well as exports suffer from the problem of under-estimation. Official statistics do not reflect the actual output of the mineral sector due to a number of factors, chief among them being smuggling and under-declaration of purchases and sales by mineral dealers, particularly after the withdrawal of designated banks from the market. This is mainly the result of both policy failure and implementation failure on the part of the government. The policy and legal framework contain loopholes, which encourage different players to operate illegally. For instance, sub-section (3) of section 73 through sub-section 2 of section 8 of the 1998 Mining Act state very clearly that no dealer's licence may be granted to an individual (or partnership) unless the individual is a citizen of Tanzania or the partnership is composed exclusively of citizens of Tanzania (URT, 1998b). Obviously, this condition is meant to ensure that the benefits from the mineral sector are internalised, among other objectives. This is certainly an important condition to discourage smuggling and promote compliance. However, one must bear in mind the fact that such a useful legal barrier does not benefit the public when its enforcement is poor. In other words, one would wish to see the mechanisms or criteria to govern its enforcement put in place.

What happens in the mineral sector today and what obviously encourages smuggling, tax evasion and cheating is that many foreigners invest heavily not only in master dealers, but also lapidary works. It is alleged that there are many foreigners running mineral dealer businesses in Tanzania, particularly in Arusha. Many of these are Kenyans, Sri-Lankans, Thais and Indians. They are said to operate their businesses using local names and in most cases local mineral dealers are involved in the conspiracy. One mineral dealer based in Arusha wrapped it up all saying:

> "Many dealers including those who are in leadership of our associations such as TAMIDA are forging alliances with foreign investors which enable foreigners to open up new investments in marketing, lapidary and mining using local names thus creating a corridor for illegal operations. These arrangements are very beneficial to the local partners as most of them retain 10% of the value added thus complicating even further the efforts to fight against smuggling, non-compliance and under-declaration of returns."

Related to this problem is the fact that some of the gems, e.g. Tanzanite, lose their Tanzanian origin and identity because they are identified with countries other than Tanzania. Tanzania is the only source of tanzanite in the world. However, most of it is exported through a neighbouring country. Many foreign countries have also used tanzanite from Tanzania to advertise their international trade thus improving their image and reputation. For this reason an official with Tanzania Mineral Dealers Association (TAMIDA) once suggested that:

> Since Arusha is the only source of tanzanite in the world, Tanzania needs to introduce certificates of origin such as "Tanzanite Trade Certificates" and make it a prerequisite that no-one should trade in tanzanite without the certificate. Also important is to introduce a fee and the income generated can be used to improve the infrastructure such as roads, schools, hospitals and water and combat smuggling. Tanzania must retain its reputation. These

> certificates could, for example, bear the country's national flag and as a long-term plan a secondary school and/or a university could be established in Tanzania (particularly in Arusha region) under the name "Tanzanite Secondary School and Tanzanite University respectively."

Tanzania could have used this opportunity to dominate and control the world market in tanzanite trade. There is an urgent need to consider the tanzanite business as a national treasure, store it as precious and protect it for the national interest. Tanzania is not going to be the first nation to protect and preserve her natural resources for the betterment of the future. Many countries have adopted similar measures.

Mineral Marketing, Price Setting and Trend

The world market price of gold began to rise in 1971 following a decline in US gold stocks, which caused loss of confidence in the dollar. A surge of speculation on the free gold market is also said to have had led to an upward trend in the world market gold price. Subsequently, the gold price went up from US $ 49.25 per ounce in 1972 to US $ 132 per ounce in 1973 as a result of the 1973 international financial chaos, among other causes. Thereafter, it fluctuated inversely with movements in the US dollar value. The price rose dramatically to US $ 507 per ounce in 1982 before declining to US $ 300 in 1985. By 1986 the price was US $ 400 per ounce and since then it has been fluctuating between US $ 300 and US $ 400 per ounce. During the months of March and April 1999, for example, the price declined from almost US $ 300 to US $ 257 per ounce following The Bank of England's decision to auction 25 tonnes of Britain's gold reserves in the government's effort to raise cash (Business Times 1999a). The price might go down further as Britain's government intends eventually to dispose of over 400 tonnes over several years. There are fears that thousands of jobs will be lost.

Prior to the macroeconomic reforms domestic producer prices in Tanzania used to be almost stagnant, despite the fact that the world market prices were fluctuating. These fluctuations could not be reflected in the domestic markets due to state monopoly. Tanzania Gemstone Industry (TGI), which used to be under State Mining Corporation (STAMICO), was entrusted to oversee operations of the entire gemstone industry, thus prices of gemstones were fixed by TGI. Similarly, prices for other minerals such as gold used to be fixed by STAMICO. Domestic prices were not based on world market price trends but rather on covering the costs of state owned marketing institutions such as STAMICO and TGI, which used the price differential for this purpose. Due to this rigid policy, prices were held constant for longer periods and Tanzania's producers are said to have lost a lot of income as a result. For example, in 1989 1 gram of tanzanite sold in Tanzania was fetching TShs 10,000, while one sold in Kenya used to fetch TShs 70,000. For many years price differentials between Kenya and Tanzania prompted illegal export of Tanzania's minerals.

With liberalisation, prices were deregulated and local mineral prices are no longer fixed. They change depending on the market situation. As noted earlier, *Meremeta* Company Limited is one of the gold marketing companies in Geita, which has been monitoring closely, on a daily basis, price changes in the world market. Since their establishment they claim to have been using prices which are closer to world market prices.[18]

The marketing structure is such that miners, i.e. claim holders and mining workers, sell their products to brokers and to a lesser extent direct to mineral dealers in the townships of the mining areas namely, Arusha, Geita and Tanga. Most miners avoid selling direct to the mineral dealers because it involves travelling to towns which costs time and money, in addition to the risks involved travelling longer distances with valuable stones. Thus they are forced to dispose of their minerals to unregistered brokers who flock to the mining villages in large numbers. Brokers normally take advantage

[18] This is based on our discussions with the General Manager of MEREMETA, January 1999.

of the market conditions to reap huge profits by paying very low prices to the primary sellers who are desperately looking for reliable markets. Existing markets are neither stable nor reliable, with imperfect information. At Umba, interviewed workers have no idea of the price prevailing in the market. Information on prices is virtually only obtained from brokers who normally distort it for profit motive.

The market is therefore biased against the mine workers and as a result this facilitates the further widening of the existing gap between higher and lower income groups in the business.

The mineral dealers' search for markets is based on contacts. They mainly sell their products abroad and therefore establish as many contacts and/or collaborations as possible with foreign marketing agents. Many such partners are located in, e.g., Germany, USA, India, Bangkok (Thailand) and Sri-Lanka. Although the number of brokers in the research areas is difficult to establish bearing in mind that most of them are unregistered, there are 79 licenced mineral dealers in Arusha, 3 dealers in Mwanza and about 5 dealers in Tanga.

Prior to 1995 gemstone prices were on the low side. The price started rising in 1996 as a result of the government's decision to close gemstone mining for one and a half months (from 15th of January to 1st of March, 1996). Immediately afterwards, the prices went down before rising again following the April 1998 accident at *Mererani*, Arusha. As noted, gold prices have been fluctuating based on the world market situation. In general, mineral prices do not reflect the social costs incurred by the miners. As we shall see in later sections, there are a number of social costs related to mining, which are not reflected in pricing.

Also important in marketing are different international mineral shows, which are either mounted outside or within the country. Mineral dealers with trade partners have also been invited to participate in international shows abroad. Others attend international shows, which are organised by TAMIDA in this country, the major objective of which is to promote Tanzania's minerals.

Hardness, colour, size and smoothness of the stones mainly determine the market value of gemstones. Tanzanite has a higher market value than sapphire partly because the former is more colourful.

Small-Scale Mining Methods and Technology

Mining methods and tools used by small-scale miners are different from those used by large-scale miners. Once the plot or claim title has been allocated to small-scale miners, a prospecting pit is excavated. In the *Mererani* Tanzanite area, pits measure approximately 1.5 by 2 metres. On average the vertical part of the pit can be as deep as 150 metres while the horizontal drift can average up to 250 metres. Small-scale Tanzanite pit excavation is carried out using simple tools such as hand hoes, spade jackhammer mattocks and drills. As the pit deepens, 50 kilogramme polyethylene bags are deployed for clearing the rubble from the pits.

Explosives are extensively used for fragmenting the rocks. Tanzanite pits become darker as they continue downward. For lighting purposes, miners use dry battery torches. The vertical pit will continue descending down to the point when the mineral veins in the rock are reached. Since veins are generally running horizontally the miners are forced to take the direction of the veins, hence the horizontal drift or heading, which was mentioned earlier. Mining requires good geological knowledge, guiding and directing miners towards the so-called 'jack-pot' where the precious stones are located. The exciting moment for a claim holder and workers is when the miners have reached the 'jack-pot'. All workers are asked to come out of the pit. Then one worker popularly known by the name 'Nobel,' flanked by one or two others will go inside the pit and initiate the blasting. The explosion involving 10 holes of dynamite can make up to 6 tonnes of rubble while explosions involving 20 holes of dynamite can make up to 12 tonnes of rubble.

Many workers have died in the mining pits during explosions. As observed earlier, before blasting, a loud announcement is made repeatedly asking workers to come out of the pits. However, there are some who choose not to come out, expecting to rush in for the kill before anyone else from outside comes to the scene. However, in most cases these people end up dead from suffocation by the smoke fumes and dust.

Gold-mining in Geita is different from the mining of gemstones at Mererani. After the potential gold area has been discovered and developed officially, the first task is to get to the ore body. This is done by digging prospecting pits, which measure approximately 1.5 by 2.5 metres. When the vertical pit is 20 metres deep, timbering begins in order to stabilise the pit sides. Timbering at *Mererani* Tanzanite mines is different from that in Geita gold mines. At *Mererani*, timbering work uses timber. They essentially make ladders (i.e. a set of rungs inserted into two long supporting wood planks) and repetitively place them inside the pit diagonally. The ladders serve as a means of ascending and descending. In Geita gold mines logs are used to plaster the pit walls to prevent the pit from collapsing. Timbering is uncommon where the rock is able to withstand the mining operations. Later on, the pits develop into a shaft which serves as a path to be used by miners for hoisting up ore as well as rubble. The shafts can be as long as 100 meters in some cases.

Tools used for pit excavation are very simple. They include picks, shovels, hammers, wedges and sharpened steel rods. Others are compressors, torches, water pumps, generators and drilling accessories. Explosive use is very rare compared with Mererani. Hoisting of rubble is done using simple technology whereby V-like timber posts are erected on both sides of the shaft. A timber log on which sisal or tree-bark rope is wound is placed in between the two poles. The log is then turned manually to elevate or lower the rope which is fitted to a leather sack for hoisting up ore and rubble or to a 25-50 litre bucket for mine drainage.

During production, the stones are collected and placed outside the pit before taking them to the crusher. Primary crushing is where the gold recovery process starts. The basic ore-processing steps in ascending order are primary crushing, screening, secondary crushing, grinding and repeated screening, pulping in a pan, amalgamation with mercury, amalgam heating and finally gold recovery.

At Umba, most of the small-scale miners share similarities with respect to gemstone mining methods. Gemstone-mining is mainly from alluvial deposits (secondary ores). Secondary ore deposits can easily be accessed compared with primary or reef mineral deposits. Excavation is carried out using simple equipment such as short-handled hoes, hammers, sharpened steel rods, spades, pan and equipment used for screening. The basic alluvial mining steps include digging a pit, collection of gravel, screening and sorting. The pits are very shallow, measuring roughly 4 by 2 by 1.5 metres. Timbering is uncommon. Large-scale mining (mining companies) have more sophisticated technology to undertake both explorations (prospecting) as well as the actual mining.

From the above discussion it is apparent that the technology used by small-scale miners is rudimentary and potentially hazardous to the miners' health and lives in general. It is obvious that for more efficient operation and productivity better methods and technology have to be sought.

4

MINING DEVELOPMENT AND THE ENVIRONMENT

The discussion and presentation of results in this section is intended to show whether the hypothesis that more environmental degradation is taking place as a result of expansion in the mining activities holds. The findings regarding movements of respondents into mining areas suggest that, during the period of economic reform (1985 and onwards), more people moved into mining areas in search of better opportunities of earning income.

In some cases, the shedding of labour from formal but inefficient enterprises and particularly mines and processing plants brought about the movement of people into mining areas. This is indicated by responses in Table 3.3 where mining companies employed 19% of the respondents before they became self-employed in the same industry. A number of factors coincided to create the "gold rush" in the early nineties. Among these are the liberalisation and permission for small-scale miners to engage in the activity of mining, retrenchment of civil servants arising from reform programmes and the dismal performance of the economy.

An observable environmental problem is the digging of pits and their abandonment as the miners move to new mining sites. This was observed in the gold mining areas of Geita (Mwanza) and Ng'ombeni and Kigwasi (Umba) where alluvial mining is practised. It was found that no refilling of these pits is ever done by artisanal miners. High cost and low enforcement of regulations are the main reasons behind this behaviour. The only attempt to refill mining pits was observed at the site of the large-scale mining company, AAPS, at Umba. According to an earlier study (Hangi, 1996), the number of pits in Mererani was estimated at 459, out of which only 100 were active. The remaining 359 were still unfilled. In Mugusu, 700 pits were estimated out of which only 300 were active. This means that 400 pits are inactive and remained unfilled.

In addition, although the effect of excavation is very location-specific, this problem is becoming acute. It is reported that approximately 4,000 ha of potentially arable land have been destroyed by small-scale gold miners in Songwa Ward in Shinyanga Region, thus causing a shortage of land for farming and livestock keeping.[19] According to this paper, land was full of open pits and cobwebs, thus unfit for any productive activity.

Another environmental problem is mining on riverbeds as observed in Kigwasi (Tanga) and in Mugusu (Geita). This could lead to weakening of embankments, collapsing of riverbanks, thus exacerbating erosion and siltation downstream. Furthermore, the situation of mineral processing activities near rivers (e.g. sluicing and washing) has consequences during the rainy season, when rivers swell and wash away the tailings and remains of mercury which are then transported downstream. This may affect water users and change the habitat for flora and fauna. During the dry season, water from streams and rivers is recycled as much as possible through holding sumps. The key issue of concern in the case of gold mining is the subsequent accumulation in sediments of mercury and its bio-magnification in the food chain within the aquatic ecosystem. The results of laboratory tests with respect to concentration of mercury in human beings, surface water and soils are available and suggest that this is a major problem (Department of Geology, 1997; DHV Consultants, 1998).

A study conducted by experts from the Geology Department of the University of Dar es Salaam revealed that most of the dumping of the heavy metal occurs in gold fields near Lake Victoria where mercury is widely used in gold recovery.[20] According to this study, uncontrolled use of mercury has led to high concentrations in various areas and poses a

---

[19] Daily Mail No. 279 of Wednesday, the 30th of December, 1998, p.1.

[20] See for example Sunday Observer, the 12th of October, 1997, pp. 3 and Mnali 1997.

danger to the people. In 1992, the report observes, Tanzania imported 1 tonne of mercury, but much of this was wasted because of the non-use of efficient mercury amalgams. The findings further show that concentration of mercury in water from rivers and ponds used for domestic purposes ranges between 0.01 and 6.78 parts per billion. In river sediments around Lake Victoria, mercury concentration was found to be in the range of 0.02 to 136 parts per billion. The findings confirm that there is indeed mercury contamination in the area. More than 250,000 miners have reportedly been involved in small-scale and artisanal gold mining around Lake Victoria in the last ten years.

The extent of mercury contamination is exacerbated by the fact that mercury retorts, which not only protect the health of mine workers by recovering 99% of the mercury vapour but also ensure efficient use of this expensive metal, are not used at all. A few amalgam retorts were distributed in gold-mining areas at Sekenke, Chunya and Geita, but our visits to Geita and Rwamgasa revealed that miners were aware of the existence of the retorts but they were not using them because they are not available commercially. Although a prototype mercury retort has been developed by the Institute of Production Innovation of the University of Dar es Salaam, it is yet to be produced on a commercial scale. It is on record that a local investor for the production of mercury retorts has applied for a licence but has not yet commenced production (FEMATA, 1999).

Other environmental impacts relate to disturbances in the natural environment, e.g. vegetation, contamination of fish, other flora and fauna and sensitive habitats. As regards vegetation, the issue concerns deforestation resulting from big concentrations of people in mining areas. The gold rush in Matabe and Rwamgasa has resulted in deforestation as cited earlier. People in mining settlements need to cut wood for cooking, building poles, agricultural activities and for timbering to support mining pits. However, a considerably larger volume of wood is used for house construction, firewood and charcoal. The direct impact of mining on vegetation clearance is the so-called grid-clearing by large mining companies. This has been observed at Umba. In Rwamgasa, the village authorities are requesting the government that they be allocated part of the Rwamgasa Forest Reserve for grazing purposes.

With respect to contamination of fish with methyl mercury chemicals, this study makes use of secondary materials. Again this problem cannot be generalised. High concentrations of mercury-contaminated fish were found at sites very close to the source of mercury discharge into the water. At least the experiences from Japan and the Amazon of people suffering from diseases associated with mercury concentration (Minamata) are known.[21]

Mining in sensitive habitats, particularly in protected areas, was also observed. This, indeed, is the major problem behind the delay in the registration of Mugusu village. The position of the government is that, once the villagers have sorted out their problems with the claim owner, they are supposed to move out of the forest reserve.[22]

Another issue concerns mining in Ikorongo. The government has granted a permit to a private South African Company to develop Ikorongo area for mineral exploration. Ikorongo is an area located outside Serengeti National Park near the Kenyan border on the northern tip of the Park. It is actually a wildlife corridor and is known to be very dependable as animals migrate to and from Kenya. Mining activity in Ikorongo area might interrupt the ecosystem of the Serengeti, thus impairing both the economic as well as environmental significance of the National Park.

It is reported that the developer has been required by the government to conduct an Environmental Impact Assessment (EIA) before any further development in the area can take place. Nevertheless, while EIA is regarded as a tool that can assist in identifying positive and negative effects of a project, Tanzania still lags behind in terms of putting in place a legal framework that is supportive of EIA. More recently the parliament passed the Mining Act (1998), which includes the issue of EIA. However, there is as yet inadequate local capacity to support EIAs, and it is not clear whether the few EIAs that have been carried out in Tanzania have helped decision-makers on the proposed development.

Although there is inadequate local capacity to undertake EIA, some foreign mining companies have shown a positive attitude towards EIA and

---

[21] Minamata is a disease that occurred in Japan, associated with uncontrollable shaking of muscles (muscle wasting disease) and blindness.

[22] Refer to a field report by officials of the Forest and Bee-keeping Division, November, 1998.

have attempted to include local communities. For example, as part of the information-sharing and public involvement in the EIA, Ashanti Goldfields company held a meeting with the Geita District Council in May 1998. In the meeting councillors raised their concerns on the following issues.

- Whether EIA procedure of NEMC is being followed;
- Whether there will be an alternative water supply so that the Geita water supply will not be adversely affected by mining operations which require a lot of water;
- Whether the open-cast method being proposed is environmentally friendly enough and if not whether there are any alternatives. Whether the refilling of the pits is possible;
- The future of the people who are going to be relocated, e.g. Mtakuja, Nyamalembo and Katoma. Where they are to be relocated and the compensation procedures to be followed;
- What will happen when the mine is closed;
- The impact of cyanide which will be in the tailings;
- Impact of trench mining; and
- Whether or not Ashanti Goldfield will restore indigenous vegetation.
- To what extent peoples' activities will be affected during the prospecting stage.

These are council issues, which address the environmental and social problems of mining.

Mining and Health Issues

Another major concern is the prevalence of environmental hazards and diseases to which people in mining communities are exposed. In Mererani, there are many cases of exposure to graphite powder and tailings that causes lung diseases. Discussions with medical and health personnel reveal the extent of this problem. In Mugusu, miners were infected by a strange disease affecting fingers and toes (locally known as *Nzoka ya Lusu*), for which no treatment is available so far. In addition, in the same area, a disease of swelling of testicles is common, but at least one can be cured.

Miners and other people in mining sites are also exposed to occupational hazards such as accidents caused by underground blasting, falling rocks due to weakened structures, inadequate ventilation and unsafe use of mercury particularly when inhaled as gas. Another danger occurs when women dry-winnow gold-nuggets and soil. In this process they are exposed to silicon dust that is carcinogenic. All these hazards and diseases occur in situations where medical services are very poor or lacking altogether. In Mererani, it was revealed that private practitioners open up dispensaries and pharmacies in mining areas just for the sake of money. The quality of services they provide is not considered seriously by the authorities. In most of the centres only unqualified staff are on duty occasionally visited by their superiors.[23]

In addition, the phenomenon of gold rush results in the springing up of informal villages which do not have basic sanitary facilities and which often bring about problems of law and order. The issue of concern here is the need to have a comprehensive strategy and programme for the provision of basic services for mining communities. The case of Mererani is one to learn from. First, there is a proper village government as well as a security committee for the mining sites. They have local militia to help with law and order. In August, 1988, Mererani had electricity installed through the contribution and efforts of the Mererani community. This is a grass-roots approach of stakeholder participation to achieve sustainable development.

---

[23] A miner narrated an incidence where he complained of stomach ache and was asked to bring in a stool for examination. However, instead of a stool, the miner brought some avocado stuff just to test whether or not these medical staff adhere to their professional ethics. The miner was surprised to be told that he had typhoid. Thereafter a quarrel between the mine worker and the medical staff ensued. Although the sick man might not have been suffering from the said illness, it is also possible that the avocado concoction might have been contaminated by typhoid because of bacteria in the dirty water.

5

SOCIAL DIMENSIONS OF MINING

Mining areas attract large numbers of people who settle in an unplanned manner. On the positive side, it can be said that social integration is achieved in terms of people from diverse cultures and backgrounds learning to live together. Indeed areas like Mererani, Kalalani and Rwamgasa are culture-rich placed in terms of the diversity of ethnic composition. As people live together, there are social benefits of agglomeration which allows everyone to learn and to benefit from each other, creating a dynamic system in which these communities conduct their daily activities and lives in general. However, in such communities there are also conflicts. This section discusses the social dimensions of mining and mining communities.

Delay in Processing Applications

A number of disputes arise resulting from the existing legal and policy framework for the mineral sector. While the small-scale miners' claims are processed at the zonal office, prospecting licences and mining licences are processed at the ministry headquarters in Dar es Salaam, without involving adequately the regional mining officers or the zonal mining officers. At zonal as well as regional levels, copies of the mineral rights issued to large-scale miners are sent after completion of the whole process for information. There are many conflicts caused by inefficiency in enforcing the stipulated law and policy directives. In addition, negligence in accomplishing and processing applications for mining rights leads to social disputes and tenure insecurities. Individuals who have submitted their applications but have not been given any reply are left in suspense and cannot begin any development until they have the papers. This delay is socially undesirable.

Acquisition of Mineral Rights

Both small-scale and large-scale miners' licence application procedures have been simplified, particularly since the 1997/98 financial year. Likewise, the mineral dealers' and lapidary licensing procedures have now become far easier than they used to be in the past. Table 5.1, for example, shows mineral dealers' reported time spent in acquiring their licences. All applications made before 1997 took between 3 to 24 months to process. Out of 15 interviewed dealers, only those who applied in 1997 obtained their licences between 1 and 2 months. It is reported that at present one would need only a month to process a master dealer licence.

The simplification of miners' licence application procedures has eased tension among mineral dealers and reduced the possibility for corruption in licence applications. This is a positive social effect of the reform programme in the mining sector.

Conflicts Related to Encroachment

This problem was mainly noted at Mererani where large-scale miners are located separately from small-scale miners. This structure has significantly minimised the conflicts between the two categories of miners. However, within blocks B and D, which are occupied by small-scale miners, disputes over property ownership are still very common. The standard size of a single claim holding is about 900 square feet or 300 square metres. This is a spacious plot, which can accommodate 3 vertical pits for oxygen, hoisting and descending.

However, in practice, there are hundreds of plots of 50 square metres and on some occasions even a plot of 10 square metres, with all of the respective claim holders possessing approved title deeds and/or mining certificates, implying that this was sanctioned by the government. A total of 304 title deeds with 50 square metre plots were approved by thegovernment in block B, where allegedly illegal miners used to operate prior to the 1996/97 financial year (URT 1997b), following demands

Table 5.1: *Date of Establishment, Type of License, Time Spent in Acquiring the License and Functions of Mineral Dealers*

S/N	Dealer's Name	Date of Establishment	Type of Licence	Time Spent to Acquire a Master Dealer licence	Functions		Employment (Lapidary)			Other Mining Rights
					Marketing	Lapidary	Lo	Fo	To	
1.	Classic Gems Ltd (A)	1990	Master Dealer & Lapidary	6 Months	Y	Y	7	3	10	Finance Mining
2.	Hai Investment Co Ltd (A)	1990	Master Dealer & Lapidary	6 Months	Y	Y	4	1	5	Claim Holder
3.	Malkay Tanzania Ltd (A)	1989	Master Dealer	6 Months	Y	N	7	0	7	Finance Mining
4.	Gems of Zodiac Co (T) Ltd (A)	1995	Master Dealer	4 Months	Y	N	3	3	6	Finance Mining
5.	International Mineral Co Ltd (A)	1995	Master Dealer	24 Months*	Y	N	4	1	5	None
6.	Glitter Gems Co Ltd (A)	1990	Master Dealer & Lapidary	6 Months	Y	Y	18	1	19	Claim Holder
7.	Colour Gem Investment Ltd (A)	1993	Master Dealer & Lapidary	12 Months*	Y	Y	8	2	10	Finance Mining
8.	Mihir Mineral Dealers (A)	1995	Master Dealer	4 Months	Y	N	4	0	4	None
9.	Blue Rocks Co Ltd (A)	1995	Master Dealer & Lapidary	6 Months	Y	Y	5	0	5	Claim Holder
10.	Arusha Mineral Centre (A)	1993	Master Dealer	3 Months	Y	N	3	0	3	Claim Holder
11.	Suchack (A)	1996	Master Dealer	4 Months	Y	N	4	0	4	None
12.	Meremeta Ltd (M)	1997	Master Dealer	2 Months*	Y	Y	123	9	132	None
13.	Tanzania Direct Minerals (T)	1996	Master Dealer	4 Months	Y	N	3	1	4	None
14.	Rocks & Crystals Min Co Ltd (T)	1990	Master Dealer & Lapidary	6 Months	Y	N	5	0	5	Claim Holder
15.	Umba Sapphire (T)	1997	Master Dealer	1 Month	Y	N	2	0	2	Claim Holder

Note: A star denotes a Joint Venture i.e. Tanzanian and Foreign Company, A = Arusha, M = Mwanza, Y=Yes, N=No, T = Tanga, Lo = Locals, Fo = Foreigners and To = Total

Source: Survey Data (1998)

and therefore pressure from the miners. Consequently, blocks B and D have, since then, been very crowded. Indeed, congestion was one of the factors which resulted in high fatalities during the April 1998 accident in Mererani.

The fact that the plots are not spacious enough to allow more than one vertical pit connected to a drift or heading, means that all the major operations such as supply of oxygen, communication and other movements are performed through a single shaft. Vital services such as communication are curtailed, especially where transmission involves sending workers up and down to deliver messages. This congestion is a potential health hazard and cause for anxiety among the miners. As one of the measures and/or conditions put forward by the government after the 1998 accident, each owner of the mining pit must install radio call for communication. Many of the claim holders are subsequently using radio calls at present.
Many drifts or headings are too long so that they are overlapping other claim titles. This problem is very common at Mererani, and it normally happens when miners are tracing underground veins which run beyond their boundaries. People have lost lives at Mererani due to blasting, which is done in one pit that is too close to another. To a large extent this is because the holdings are small and too close to each other. Approximately 75% of the 1998 conflicts recorded by the Mererani village government are related to land and mining rights as indicated in Table 5.2.

Table 5.2: *Estimated Number of Disputes as Recorded by Village Leaders in 1998*

	Village Name	*Average Number of Cases Reported per Day*	*Number of Disputes per year*	*Disputes Relate to Land and Mining Rights (%)*	*Other Disputes (%)*
1.	Mererani	6	2190	75	25
2.	Rwamgasa	4	1460	85	15
3.	Mugusu	1	365	75	25
4.	Ng'ombeni	0.25	91	50	50
5.	Kalalani	2	730	60	40
6.	Kigwasi	2	730	75	25
7.	Mwakijembe	0.14	52	50	50

Source: Survey Data (1998)

The seriousness of such disputes has prompted the Mererani village government to create mines security committees, which are entrusted to verify and resolve related conflicts. Quarrels over veins inside the mining pits can sometimes be very serious to the extent of causing injuries and even loss of life.

Table 5.3 provides statistics on the major problems affecting miners' work performance. As can be seen from the table, about 38% of the respondents rank "poor equipment" as the first major constraint while 12% rank it a second major problem. This problem is more acute in Geita where 56% of the respondents revealed this experience. "Injuries" are major problems with approximately 17% of the respondents ranking it the first major problem. About 19% of the sample rank "injuries" the second major problem. Many of these problems happen frequently and most injuries are caused by accidents, which are a result of negligence, congestion and conflicts.

Table 4.3: *Major Problems in the Mining Activities in Tanzania, 1998*

Sn.	*Problem*	*First Major Problem*	*Second Major Problem*
1.	Poor equipment	37.9	12.07
2.	Injuries	17.2	18.6
3.	Limited Capital	10.3	16.9
4.	Infected Diseases and Health hazards	8.0	10.6
5.	Lack of clean air	5.7	20.3
6.	Lack of water	3.4	5.7
7.	Other	17.5	15.9
8.	Total	100	100

Source: Survey Data (1998)

Encroachment related disputes are very common. The creation of mines security committees, for example, is a necessary solution but it does not seem to be sufficient. If anything, it actually fuels the chronic problem of corruption.

Conflicts Between Mining Companies and Claim Holders

In view of the positive contribution of small-scale artisanal miners to the economy and as one of the ways of combating poverty, the government is committed to supporting the small-scale mining sub-sector by facilitating the transformation of the present small-scale mining approach into a more organised and modernised one. Among the six strategies for rationalising small-scale artisanal miners include promoting partnerships and joint ventures between local small-scale miners and large scale investors to facilitate technology transfer and optimise mineral resource exploitation (URT 1997a; URT 1997b). This was a government's move to persuade small-scale miners to accept joint ventures or partnership with large-scale miners on the pretext that small-scale miners have low capacity to operate efficiently.

During the course of implementation of the policy, district level mineral officials are alleged to have used force in some areas such as Geita to push small-scale miners to accept and sign vague agreements with large-scale miners. One such allegation is from Buck Reef Gold Mines, claim title no. 40286 in Geita. The owner of this claim noted that claim holders faced pressure from government officials forcing them to enter joint venture or partnership with mining companies which are surrounding them. The small-scale miners were sometimes threatened by the government with the revocation of their title deeds if they did not abide by the directive. As a result of this pressure a number of small-scale miners lost their property by being bought off by mining companies, many of them at throw-away prices. Others signed very conditional agreements unknowingly. In fact, many of them have signed contracts with mining companies without knowing what exactly the content of the document was.

Mutagwaba, *et. al.* (1997) assert that during the on-going liberalisation process where small-scale miners enter into negotiations with bigger companies interested in some of the claim holdings, the idea of fair play is still far-fetched. While mining companies are in a position to engage lawyers, engineers, economists and other professionals in negotiations, most small-scale miners do not have that capacity and can hardly understand, interpret or sometimes even read the negotiated contracts. Many such cases can be found at Mugusu, Rwamgasa, Nyakagwe, Nyarugusu and Nyamtondo in Geita. For some time now, pressure has been applied by both mining officials and the mining companies to buy off the small-scale-miners and/or subject them to sign the agreements, which are not broad based.[24] There are, therefore, many disputes which are related to irregularities in drafting and signing such contracts. Many of the victims are claiming that they have been swindled out of their property by mining companies.

The outcome of the takeover by large-scale miners has been a serious problem of subordination of small-scale miners to conditions imposed by large-scale miners. In addition, since most large-scale miners are foreigners, this takeover implies the displacement of local people from the ownership of the resources, thus fostering disputes between foreigners and local entrepreneurs. The fact that this is done with the backing of the government means that small-scale miners are not given due support as the policy stipulates. Because prospecting licences involve large areas, many of which not only enclose claim holdings but also employ sophisticated machines, small-scale miners are indirectly humiliated and sometimes forced to accept disproportionate compensation and leave the place.

Table 5.2 revealed that 85% and 75% of the disputes, recorded by the Rwamgasa and Mugusu village leaderships respectively, are related to land and mining rights. This number is said to be on the lower side because people have lost confidence in their village governments when it comes to reporting disputes involving mining companies. Village governments have not been empowered to resolve such disputes, and sometimes corruption has been cited as the other limiting factor.

Conflicts Between Claim Holders and Mining Workers (Village Communities)

In November 1997, *Meremeta* company, involving the government of Tanzania and ruling companies of South Africa, was created. The role of *Meremeta* is to provide modern mining equipment for small-scale miners

[24] Normally, the mining companies revise the contracts in their favour especially after a certain period of prospecting, i.e. when they realise that the claim holding does not appear to be as rich in deposits as they had anticipated. Many of the revised contracts have tended to reduce the benefits destined to claim holders in the previous version of the contract.

and a market at competitive prices. The equipment includes water pumps, compressors, crushers and jackhammers (URT 1998c), supplied on the condition that the minerals are sold to *Meremeta.* Under a different agreement the village governments also acquire crushers, provided by *Meremeta,* and administer them. This was evident at Mugusu and Rwamgasa villages. At Mugusu, the village government owns two crushers. The village meets the cost of fuel and *Meremeta* pays for maintenance. The village government charges Tshs 2,000 per 50 kilogramme bag of stones and an average of 40 bags of stones are crushed per day which earns the village a daily income of TShs 80,000. The net profit margin per day is approximately 20,000 TShs. Private crushers are also available at a charge of Tshs 4,000 per 50 kilogrammes of stones.

The system of providing equipment has made both mining activities and mineral processing relatively easy and many interviewed mining workers and village leaders appeared to be very optimistic. Production has increased because, unlike before, with equipment like water pumps, it takes the workers a very short time to drain water from the pits (water drainage) and start production. Likewise, with Meremeta's crushers it takes them a relatively short time to carry out benefication process. Furthermore, employment is said to have increased tremendously in the mining communities due to the availability of working equipment. As noted, when equipment like compressors, water pumps and crushers are available production stoppages are very rare and therefore workers are busy throughout, with incomes likely to increase.

Similarly, when production of minerals is higher many other auxiliary activities also mushroom thus creating employment opportunities in the mining communities.

However, a number of obstacles which interfere with Meremeta's programmes to assist the miners has emerged. The decision by some of the small-scale miners to sign new contracts with mining companies have caused havoc to mining workers as well as village communities, following the harsh conditions attached to the agreements, which are also eroding the benefits which society used to reap. Some of the contracts with mining companies prohibit small-scale miners to enter into any other agreement with a third party.

Another problem is that small-scale miners are also prohibited from using any sophisticated equipment like water pumps, air compressors, etc. thus freezing the employment of many people and denying the small-scale miners access to new technology. Mining companies interviewed, e.g. Ashanti Gold Fields, Pangea and East African Mines, acknowledged and confirmed such a tendency, but argued that they will be the losers if they did not impose restrictions. For example if they allow the small-scale miners to employ sophisticated machines, all the mineral reserves around that area will be exploited (depleted) before they finish prospecting.

At the time of the survey, mining workers at Mugusu and Rwamgasa were not working because production had stopped following withdrawal of Meremeta's working gear. Indeed, shortage of equipment is one of the critical problems which has impacted negatively on production and employment in the mining areas.

In addition, there is limited job security for the mining workers when their employer, i.e. a claim holder, signs a contract with mining companies without involving them. Indeed, what is emerging here is the marginalisation of mining workers.

Claim holders try to maximise benefits, because what matters to them is the compensation they get from mining companies. They do not pay enough attention to what will happen to the mining workers and/or the village community when they sign such contracts. Virtually all these developments have created many disputes, hatred and conflicts between villagers and mining workers on one side and the claim holders and mining companies on the other. At Mugusu, the life of claim holders who have signed such contracts with mining companies is at stake. Claim holders live in fear and therefore in tension. For both mining companies and small-scale miners, this is not a friendly co-existence, which they badly need for smooth execution of the mining operations.

Land Use Disputes

Land ownership is the most important factor in the development of small-scale artisanal mining in Tanzania. Confrontation between the people in most of the areas where small-scale mining is practised is very common.

The most conspicuous land dispute is related to interpretation of land rights particularly in terms of compensation. Also important are conflicts arising as a result of mining rights overriding land ownership rights. The findings reveal that there are two major causes of land conflicts in the mining communities.

The first factor is lack of planning and co-ordination at national level. It is very common to get a prospecting licence attached to a map showing a smaller number of existing properties than the actual number of properties at the site, i.e. there are more properties than those indicated on the map issued to mineral investors. Villages, hospitals, schools and claim holdings are some of the properties that may not be reflected in the map. It therefore becomes difficult for the investors to resolve all the disputes arising afterwards, mainly because the claimants and/or owners of these properties become resistant and uncompromising. They tend to believe that the mining companies have cheated and swindled them out of their properties. Likewise, following the number and types of properties indicated on the official map, the mining companies do not feel they have any obligation to compromise. Thus, the situation becomes tense, fuelling more disputes in the villages. Table 5.2 shows that land and mining disputes recorded by village leaders dominate other types of disputes, claiming between 50% and 85% of the total number of disputes in the sampled villages.

Related to the above problem is the fact that some mining companies have been allocated prospecting licences within a reserved or gazetted forest or a village. Some of these villages are said to have been making unjustified demands on the mining companies and as a result these companies have ended up rejecting such claims thus creating a hostile environment in the neighbourhood. This is a common problem in Geita and is the result of inadequate capacity, which is further complicated by the absence of an efficient system to update the baseline information.[25]

[25] Many properties in the countryside are not known (registered) by the central government. For example, there are many villages, which have been existing for, but they are not registered yet. Many such villages applied for registration a long time ago but the government has not processed the applications.

Indeed the result of our study confirms the point that inadequate involvement of the intended people fuels disputes in the village communities where mining is carried out. Table 5.4 shows registration status of the sampled mining villages.

As revealed by the table 4 villages out of 7 have not been registered yet, despite making their applications a long time ago. One of the reasons for the delay is that some of the villages, e.g. Mugusu in Geita, Kigwasi and Mwakijembe at Umba, are located in Geita and Mkomazi Forest Reserves respectively. The properties in these villages cannot be reflected in the national records thus complicating even further the issue of fair compensation for claim holders.

Table 5.4: *Registration Status of the Sampled Mining Villages in Tanzania, 1998*

Sn	*Village Name*	*Population Size*	*Year of Application*	*Current Status*	*Registration No.*
1.	Mererani	40,000	1975	Registered in 1976	-
2.	Rwamgasa	25,000	1975	Registered in 1976	1977
3.	Mugusu	15,000	1988	Not Registered	-
4.	Ng'ombeni	900	1989	Not Registered	-
5.	Kalalani	12,500	1989	Registered in 1989	TA/K/551
6.	Kigwasi	2,000	1989	Not Registered	-
7.	Mwakijembe	700	1989	Not Registered	-

Source: Survey Data (1998)

Second is the misconception on the part of the indigenous people of legal rights over mining as well as land ownership as stipulated in the law. Ashanti Gold Fields in Geita has been receiving many compensation claims from people who had already been compensated in the past. Most of them are trying to make money out of what they believe is an attractive opportunity. In fact, people within the licensed areas are still constructing houses or selling their properties to other individuals knowing full well that the ownership of the properties has already been transferred to the mining companies. Ignorance is rampant and may lead to more social problems and conflicts.

Conflicts Between Smallholders and Claim Holders

There are many disputes between claim holders and smallholder farmers especially at Rwamgasa in Geita. Farmers plant crops, sometimes perennial crops, in areas with title deeds. Claim holders are advised by the mines' officials to allow smallholder farmers to plant seasonal crops in areas which are not operating temporarily, but farmers plant tree crops instead. Both annual and perennial crops are tricky, because when the owner intends to start production, farmers owning the crops demand either joint ownership of the area or unjustified huge sums of money from the claim holder as compensation. Such problems are even more complicated when the crops were planted by people other than the current farmers without clarifying the property rights. This has also created many conflicts in Geita and points to a weakness in the system of allocating plots.

The 1979 Mining Act stipulates that allocation of a title deed in the area with farms, trees, graveyards, etc., will require compensation from a claim holder by mutual agreement between the two parties. When such an agreement is futile, it is possible to refer these cases to the Minister. In practice, many of the claim holders prefer partnerships whereby 40% of the total shares are allocated to the claim holder, 10% to the owner of the farm and 50% to the mining workers although, officially, all the minerals belong to the claim holder.

This is a system which has been developed locally. It is reported that such informal agreements are very effective in terms of minimising land disputes. Official compensation processes are not popular because land officials use official rates which are outdated and very low. Also the litigation process is costly in terms of time and money since most courts are far away from the mining sites.

Some conflicts are historical in nature because most of the mining areas in Geita were surveyed in 1939 during the British colonial era. Some of the present residents came to seek employment with foreign companies which used to mine in the areas, thus residing in these villages. They started farming at a later stage when production of minerals by the companies stopped. When foreign mining companies closed down their businesses, prospecting rights were granted to newcomers as most indigenous people were not interested, reportedly, because mining was considered to be an inferior occupation. By then small-scale artisanal miners were popularly known by the local name "Wachoji," literally translated as "prospector" or "explorer." Today many of the indigenous people are searching for mining rights in the areas, which are already allocated to others. They go even further and accuse the authorities, mining companies as well as small-scale miners of denying them their mineral rights. Many indigenous people believe that they have been marginalised because they were the first people to settle in the area. The fact is that they have now realised that a title deed is a valuable asset, which can make a lot of difference between being rich and poor. What is emerging here is that the failure of the earlier settlers to obtain title deeds and to realise the importance of mining is costing them dear about 60 years later. This is a case of missed opportunity.

The Quality and Provision of Social Services

There is a general assumption that mining development is likely to spur improvement in the delivery of social services. However, this was not the case in the study areas. Research findings indicate that there is poor provision of essential services in all the mining sites surveyed.

Water Supply

Water supply is a problem in all three areas. In Mererani, water is collected from a river some kilometres away, on donkeys and carts. In Mugusu, water is drawn from the river nearby, where gold processing is also taking place. In Kalalani, water is from wells and is very salty. In Kigwasi, people use water from Umba River directly. Therefore diseases caused by poor sanitation and unsafe water (typhoid, diarrhoea, etc.) are common. This was also confirmed by health personnel stationed in these areas.

Health Issues

The problem of diseases and inadequate medical services is a serious one. There is limited awareness about diseases, especially HIV/AIDS and sexually transmitted diseases (STDs), due to a general "care-free-attitude" that is typical of mining communities. There is a feeling that, because work is so hard, one needs to enjoy oneself when one earns money. In Mererani, there are at least four private dispensaries, including one belonging to an international NGO (Marie Stoppes). In Mugusu, there are two private dispensaries and occasionally the ward health officer visits the village to offer services. The situation in Umba is not very different. In Kalalani, there is a government health centre, while in Kigwasi there is none. The people of Kigwasi used to get medical services from AAPS, but since relations have soured, that is no longer the case. Indeed, AAPS had to close its health facility due to bad relations with the local community. Workers who need medical attention are taken to Kalalani, Korogwe or Tanga. One may wonder why such a place like a mining community, where income generation goes on night and day at a relatively high rate compared with other more sedentary communities, should have such a high rate of disease incidence. The explanation lies in the nature of how the 'townships' or communities come into being! The 'gold rush' phenomenon brings people from all corners of the country and the world at large. Their main objective is to get the minerals and invest elsewhere, giving a temporary feel to miners' lives in the communities.

Since the mining areas are normally located in the wilderness, no planning for the communities ever takes place prior to the growth of the normally makeshift settlements. This facilitates the haphazard nature and congestion of buildings where most of the basic amenities are lacking, such as, clean and safe water, sewerage and drainage systems, toilets and so on. However, some of the diseases are occupational hazards.

Child Labour

Another social problem common in mining communities is that of two kinds of child labour. There are the school children who are attracted to working in the mines to earn an income or to help their parents, causing them to abandon school and go to the mining areas in search of employment. In most cases, such children are pushed into this situation by poor relatives or parents who use them to generate income. Poverty in the households is the main issue. This is interesting because these areas are relatively wealthy through the minerals that are obtained. However, poverty is a serious problem as reflected by simple indicators such as quality of social services. It is only to be expected however because wealth is concentrated in a few hands and there are no equitable distribution mechanisms. The government has taken some measures to discourage child labour in the country according to the mining regulations released recently.

Box 5.1: *Ban on Child Labour in Mining*

> Employment of children in mining areas throughout the country is officially banned with effect from Wednesday, the 11th of August, 1999 according to the Mining Regulations whose enforcement starts immediately. A person holding a primary licence, shall not cause children below the age of 16 to be employed or engaged in any mining or processing operations in his primary mining licence area.
>
> *Daily Mail: Monday, August 9, 1999*

The second category is that of children who neither have parents nor go to school, i.e. street children. In the case of Mererani, the street children go there to eke out a living, where, they find ready employers who use them for various tasks and errands such as sending messages inside the mines. They are popularly known as *nyoka* (snakes) because of the ease with which they can manouvre their way around. Some of the people who employ such children argue that they are actually doing a favour to the children because, otherwise, such children would be in the streets and doing worse things than what they are doing at Mererani.

Other Social Problems

Other social problems in mining areas include use of drugs, paricularly bhang or marijuana, and alcohol. Although such habits already exist in the community, the concentration of people in mining areas, the nature of the work and the mentality prevailing there, create a conducive environment for such activities to take root in alarming dimensions. This goes hand in hand with prostitution which is rampant in most mining areas and is closely linked with the spread of diseases such as HIV/AIDS and STDs. In addition to prostitution, in such areas there are cases of rape and sexual abuse. In Mererani, cases of rape have been reported, but since there was no police post (until late 1998), no concrete measures were taken. Furthermore, there are reports of collusion between the culprits and rich individuals, who want to protect the culprits by buying off the "cases" from the victims. In such situations, victims lose their rights and become socially stigmatised. Another problem associated with mining communities is the prevalence of "strange and anti-social" behaviour and beliefs about how one can get rich. Some miners believe that rape, impregnating young girls and taking away their babies (particularly if they are male), is necessary for one to become rich. These are strange and frightening beliefs normally perpetrated by traditional medicine-men, who are found near the mining areas.

Another social problem arises due to the 'sudden death' nature of employment loss to the mining communities. This comes about as a result of joint ventures between local claim holders and foreign companies, whereby the latter imposes conditions on the local Claim Title (CT) holder to stop production while the foreign company is prospecting. This is the root cause of the problem between several CT holders in Mugusu on the one hand and mine workers and their families on the other. This is in addition to the problem of Mugusu being located in the forest reserve, from where the government is intending to remove them once they have sorted out their problems. From a social point of view, the loss of livelihood of a whole village community is imminent under such conditions.

In the case of Geita, a meeting between Ashanti Goldfied company and the villagers have revealed some social issues, which ought to be taken into consideration, including, relocation and compensation and change of socio-economic lifestyles when a big mining company starts operating. Some of these issues were raised by Mtakuja villagers and included specific aspects such as:

- Inadequate compensation for crops and other assets;
- Full payment of compensation money not reaching affected parties;
- Disturbance of burial sites;
- Valuation of land being undertaken at a season that will reflect lower values;
- Financial loss associated with re-growing of crops and the time it takes before such crops are ready for harvest;
- Relocation and the increase of distances to farms;
- Concern about relocating elderly people, especially those who are unable to rebuild their houses;
- Potential division of communities due to relocation;
- Loss of income-generating resources in the proposed mining area;
- Concern about quality of places where people will be relocated in terms of fertility of land; and
- Concern about uncertainties created by relocation, e.g. planting of crops, building of houses, etc.

It was notable also that people were concerned about other general socio-economic impacts such as:

- Whether or not villagers will benefit from the mine and to what extent;
- Potential influx of job-seekers to Geita and surrounding villages and the impact of this influx on social service provision;
- Whether locals will be given special consideration in job allocation;
- Loss of grazing land due to mining;
- Concern about potential displacement of people due to development of a dam for the purposes of the mine;
- Diversion and loss of access to public roads;
- Loss of employment for artisanal miners; and
- Impact of mine closure on the local economy.

Institutional Issues

Operation of the mineral sector involves a number of stakeholders and institutions that are also supposed to be co-ordinated for the efficient performance of the sector. Both horizontal as well as vertical institutional integration is lacking. Inefficient institutional co-ordination has adversely affected performance of the mineral sector in the country.

For example, in talking to different authorities about this problem, there appears to be limited co-ordination between the Ministry of Energy and Minerals, Ministry of Water, Ministry of Local Government and local leaders on the question of mercury contamination and its consequences. In addition, mining of gemstones in Kigwasi and Umba is taking place illegally in Mkomazi Game Reserve. Before the area was designated as a game reserve, local people used to get six-monthly mining permits. Since being gazetted into a game reserve, no more permits are being issued. This notwithstanding, locals still believe that the reserve area is very rich in gemstones, so they continue to mine illegally without the knowledge of the top authorities.

In this study, it was found that gold-mining in Mugusu is located in a forest reserve. The issue is known to officials of the Forest and Bee-Keeping Division of the Ministry of Tourism and Natural Resources.

As noted, most of the maps attached to mineral rights issued to large-scale miners do not reflect or indicate important properties located on the respective piece of land. On many occasions, there are more properties on the ground than those indicated on the map issued to mineral investors. Villages, hospitals, schools and claim holdings are some of the properties that may not be reflected on the map. This is one of the factors which has tended to cause conflicts between mining companies and indigenous people. It shows that there is no co-ordination between the higher authorities at ministerial level and the local governments which could have been utilised by the central government to update the baseline information.

However, the periods granted for reconnaissance, prospecting and renewal are rather disquieting. Section 29 of the 1998 Mining Act provides for the duration of a grant and the renewal of a licence. A period not exceeding two years is granted to the applicant for preliminary reconnaissance. Thereafter a period not exceeding three years is granted to the applicant for the initial prospecting. The prospecting licence is renewable for two successive periods. Thus, for applicants interested in undergoing preliminary reconnaissance, a total of nine years is granted before the actual production starts, whereas for those going straight to the initial prospecting, a total of seven years is granted before actual production begins.

One of the main activities during prospecting is to establish or identify mineral reserves available in the given area, an activity which allows prospectors to carry large amounts of mineral-bearing rock for analysis to laboratories overseas. Ashanti Gold Fields, for example, used to take its samples abroad, but at present it takes them to Mwanza where there are three laboratories, according to the Tanzania Chamber of Mines (TCM). This situation has fuelled suspicions that a lot of precious minerals are being siphoned off by prospecting companies mainly through the transportation of samples and other means. One cannot rule out this possibility, particularly when prospectors are granted such long periods.

There is need to question the rationale for granting such a long reconnaissance as well as prospecting periods to some of the prospectors. Given Tanzania's low capacity to monitor all these activities, one cannot rule out the possibility of illegal mineral exports through this method. The present environmental policy seems to provide private entrepreneurs unlimited freedom to operate without being monitored effectively.

The government needs to look at such problems rigorously and establish a mechanism, which will tie up all loose ends so as to ensure the local economy benefits. Respondents from mining companies and the Tanzania Chamber of Mines repeatedly defended the decision to take rock samples abroad for two major reasons. Firstly, they argue the allegations are ill-conceived because the cost of sending the rock samples would be much higher than whatever gold or diamond was likely to be contained therein. A rich mine like Kahama, with about 15 grams per tonne, would need a planeload of 10 tonnes for extracting 150 grams of gold valued at a maximum of US $ 1,500 at the optimal price of US $ 310 per troy ounce. Airlines charge about US $ 2.85 per kilogramme of cargo, making it both prohibitive and pointless to carry such loads in order to extract minerals later. Secondly, sending those samples is justified since Tanzania lacks the modern laboratory facilities needed for metallurgical analysis of mineral bearing rock.

It therefore follows that there is a great need to strengthen and improve local laboratories to meet international standards, in order to discourage further transportation of rock samples abroad. Improvement of local laboratories will, among other benefits, attract mining samples from other countries.

6

EXPERIENCES FROM OTHER COUNTRIES

SADC Countries

A report by the SADC Mining Co-ordination Committee indicates that the mining sector continues to be the backbone of most economies in the region. The average contribution to total foreign exchange earnings in the region's economy is 60%. In the past ten years, the sector contributed 10% of the GDP and 5% of formal employment. This achievement is partly a result of current member states' policies and legislative provisions, which have resulted in increased exploration. The major mining nations in SADC include Botswana, Zambia, Tanzania and Zimbabwe of which Tanzania does not feature prominently. Most of these use large-scale mining companies, which have their own marketing networks. However, there are also small-scale miners with problems similar to those discussed in this book, i.e. environmental and social problems. Among the measures taken to avoid such negative impacts include promotion of security and preventive (safety) facilities within the mining pits.

Bolivia

In Bolivia, it is estimated that about 400,000 out of the 7 million inhabitants live on small-scale mining. Their activities are reported to be causing severe environmental problems, thus affecting not only the miners themselves, but also the neighbouring communities. The most critical environmental problems caused by small-scale miners are:

- Mercury emissions in gold mining;
- Siltation of rivers caused by alluvial mining;
- Destruction of the landscape caused by alluvial mining; and
- Contamination of water caused by concentration of gold processing plants.

Efforts are being made to reduce the use of mercury by introducing better methods and techniques, also providing equipment and material for the optimisation of mercury use and minimisation of mercury contamination. Efforts are also being directed towards reducing water contamination from existing mining activities and from old tailing dams. Thus a water treatment plant has been installed.

Thailand

Thailand is a mineral-processing zone in Asia mainly processing (i.e. cutting and polishing) imported gemstones, constituting the biggest proportion of Thailand's international trade. Most of the mineral deposits in Thailand, e.g. ruby are not extracted as the country has decided to reserve it for the future generations.

Lessons for Tanzania

Evidence has shown that Tanzania has abundant mineral reserves in various areas and the number of new discoveries during the liberalisation period kept on increasing.[26] Development in the mineral sector will become meaningful only when the country reserves some of the mineral deposits for future use. The present trend of mineral extraction in the country is bordering on extravagance.

This study has not dwelt on the quantification of costs and benefits of the mining activities in the research areas. Indeed, comparison of economic benefits and costs could have provided realistic measurements in terms

[26] See for example URT (1997b), URT (1998c) and Daily News (1998).

of whether or not the mining communities benefit from mining activities. An economic and financial analysis of this kind should therefore be an area for future research.

Nevertheless, judging from the preceding qualitative discussions made during the field work and other economic, social and environmental analyses, the mining activities as presently practised at Mererani, Geita and Umba cannot be economic when all the costs in terms of the economy, society and the environment are compared with the benefits.

For example, corruption, which has tended to divert benefits from the eligible segment of the population to the ineligible, is almost chronic in the country. As reported earlier, the 1998 corruption perception index computed by Transparency International ranks Tanzania fourth out of 85. Likewise, some important human development indicators for Tanzania suggest no improvement during the reform period. For instance, life expectancy at birth has declined from 54 years in 1990 to 48 years in 1999.[27] The proportion of the total population with access to health services has also declined from an average of 81% during 1985-87 to 42%. Safe water is accessed by only 38% of the population, far below the 56% during 1985-87. Per capita daily caloric intake as a percentage of the requirement has also tended to decline from 69% in 1965 to 65% at present.

[27] UNDP (1991).

7

FOR WHOSE BENEFIT IS MINING DONE IN TANZANIA?

The previous chapters have discussed the mining industry status and its potential for the development of the country. We now turn to discussion of the substantive issues in terms of achieving the desired sustainable development of the country as contributed by the mining industry. The major question asked here would be; for whose interest and benefit is mining done in Tanzania? To many Tanzanians, the issue is not so much for whose interest because we know that minerals are part and parcel of our natural resource endowment, which are supposed to be utilized in order to provide the necessary inputs to facilitate our nations development for improving the people's welfare. However, this presupposes that the manner in which the exploitation is undertaken will determine whether the major stakeholder groups equitably share the benefits of the utilization of our natural capital. In this chapter therefore we discuss the issues pertaining to revenue accrual, its actual amount and potential, investment incentives and benefits, in country mineral processing, smuggling allegations and operational weaknesses. Furthermore, we discuss conflicts and accidents and also compensation problems of expropriated individuals in mining areas.

Box 7.1: *Some Key Facts*

- Global demand for gold was at 3,642 tonnes in 1995, as against the previous record of 3,573 tonnes in 1992.
- Gold is uniquely malleable and ductile; one ounce can be drawn into 50 miles of thin gold wire.
- More than 70 countries, including 31 in Africa, changed their laws to attract foreign companies by the late 1990s.
- South Africa is the worlds leading producer; in 1995 it produced 523 tonnes. But production is falling.
- By 1991, just over 20 companies delivered more than two-thirds of official global gold production (outside of China); the world's two largest companies, RTZ-CRA and Anglo-American, controlled nearly one half of this amount.
- Brazil is the biggest single small-scale gold producer, with annual production of 80 tonnes. Some 500,000 miners work in the produce.
- In Brazil it was found that more that 30% of miners had mercury levels above the WHO tolerable limits.
- It is estimated that small-scale miners produce up to 20% of Africa's gold.
- In South Africa, each tonne of gold mined costs one life and 12 serious injuries.
- In Peru, companies estimate that one mining job creates 10 additional jobs in the area.

Source: URL http:/www.panos.org.uk/briefing/gold.html

Investment Incentives vs. Benefit Sharing

Foreign investors look upon mining as just another global investment activity that needs to be analysed in the global investment opportunities context. Therefore, the fiscal regime of a mining regulation is an important factor that will attract foreign investment into the country. On the other hand, governments as custodians of the minerals on behalf of the people, looks upon the development of mining activities as avenues of capital accumulation for economic and social development of the country. They have the taxation powers to facilitate its developmental objectives. A major objective of governments is to recover resource rent[1] and to use it for the benefit of the country. Lange and Hassan (2000) point out that, "...from an economic perspective, sustainable and equitable management of these resources requires that the resource rent be recovered by government through appropriate taxes.[2] Furthermore, it is important for sustainable development that, some significant amount of these resource rents be re-invested in other assets like renewable natural capital assets and other economic activities, which can replace the employment and income generation of the mineral resource based industries, once these resources are exhausted. De Vera (2000) acknowledges the fact that balancing the conflicting interests of the government (attracting investors and maximising revenue at the same time) is a daunting task. Ideally, a taxation regime for mining must be low enough to attract investments both local and foreign; but high enough to ensure sufficient revenues are generated for the government.

The mining sector in Tanzania accounted for less than 3% of the total workforce and about 5% of GDP in 1997 (Coakley 1998). In October 1997, Tanzania formulated a new minerals policy giving priority to private sector initiatives, the rationalization and organization of artisan and small-scale mining and product marketing, and new initiatives to mitigate the adverse environmental and social aspects of mining activity in the country. In the same year of 1997 Financial Laws Acts provided improved fiscal incentives for the mining sector while in the same year a new investment act created the Tanzania Investment Centre. Investment in the mining sector in Tanzania thus was given a new lease of life by all these efforts to

[1] Resource rent is by definition a profit higher than the average for the economy, attributable to the scarcity of the resource relative to demand for the resource on the world market. In the case of depletable resources like minerals, Hotelling's rent is applicable, which is the area between the market price and aggregate marginal cost based on total quantity consumed in the market (Hackett 1998:313).

[2] Botswana and Namibia have been fairly successful at recovering the rent generated by mining in that taxes levied by the governments have been roughly equal to the rents generated.

lay the foundation for facilitating the mining activities to contribute to the economic development of the country. Among the major incentives provided by the investment Act was the guaranteeing of mining investments against nationalization and expropriation, provides for a 30% corporate tax rate, capital allowance deductions, depreciation and other tax incentives. Additionally, mining royalties are set at only 3% of net sales revenue and no royalties are paid on cut and polished gemstones. The later is meant to motivate investors to invest in local mineral processing which has more benefit to the local economy in terms of value added being retained locally. However, even with such incentives, only a handful of small-scale mineral processors have so far invested in mineral processing in country.

The country has successfully attracted mineral exploration and investment since 1992 which has triggered a "gold rush" in the gold bearing areas of the Lake Victoria region of Mwanza (Geita), Mara (Buhemba), Shinyanga (Kahama) and Tabora (Nzega). Gold exploration and discovery dominated the mining industry news during 1997/98, with several mining companies mostly foreign announcing the discovery of significant gold resources in Geita, Nzega, Bulyanhulu and other areas. It was estimated that by 2001, annual production of nearly 30 metric tons of gold was possible; with additional potential for growth Tanzania is destined to become one of Africa's major gold producers (Coakley, 1998). Investment in the mining sector thus increased tremendously as presented in chapter 3, Table 3.5 where it is indicated that between 1990 and 1999, a total amount of US$ 238,415 million was invested in the mining sector. The Tanzania Investment Report puts the stock of Foreign Direct Investment (FDI) in the mining and quarrying sector for 1998 and 1999 at US$ 503.6 million and US$ 848.9 million respectively, while the flow of FDI for the same sector has been put at US$ 345.3 million for 1999 alone. As shown in chapter 3 Table 3.11, gold production began rising as a result of these investments from a low 427 kilograms of gold in 1998 to 4,767 kilograms in 1999 and an even higher level of 15,060 kilograms in 2000, which boosted foreign exchange earnings to US $ 39.8 million and US$ 120.57 in 1999 and 2000 respectively (Economic survey, 2001).

The above mining development in the country has been facilitated by a concerted effort to attract investor in the mining sector through various incentives. Although providing attractive terms to foreign investors would provide for a healthy competitive position for the country, one may want to ask what should be the optimum incentive package that will attract foreign investors and yet provide the maximum benefit to the country and its people? There is a feeling in various circles that Tanzania is giving away more than necessary "investment incentives" to the loss of the country especially considering that minerals are exhaustible resources. Table 7.1 below compares various mining taxes for 23 countries as an incentive structure for investors. Also the following Table 7.2 shows the royalty rates applied by various mining countries in the developing and developed world. If one critically looks at these tables, one will easily see that Tanzania has been overly generous to investors in the mining sector compared to other countries at the expense of its people. While 35% of the 23 countries presented in Table 7.1 charge no royalties on minerals but this is more than compensated for by other means of taxes as shown in Table 7.4 below. While Tanzania charges a 3% royalty on smelter value of gold, it does not charge VAT nor import duty and interest withholding taxes! On the other hand, those countries, which do not demand royalty as an incentive to investors, do charge hefty income taxes of mostly more than the 30% Tanzania is charging. Furthermore, they make up for the royalty not demanded by charging VAT of between 7% (Canada) and 25% (Sweden). In addition, they also charge import duty of between 0.5% (Canada) and 12% (Peru) and higher dividend withholding taxes. All this goes to show that the situation could be improved for the people of Tanzania by closely competing with other mineral rich countries through adjustment of the incentives to be more favourable for the owners of the mineral resources i.e. Tanzanians.

The argument that Tanzania lacks the infrastructural services found in some other mining countries and hence has to offer more attractive incentives relatively cannot be justified by providing continuous and costly incentives. In all countries mining activities are done in areas where the economic infrastructure is not well developed. It is true that in some of the countries the services will then be put in place relatively more quickly to facilitate the mining activity than in others. Yet in some countries, Tanzania included, the investors are obliged to put in place the necessary economic infrastructures of water supply, electricity and roads among others in order to engage in their primary activity of mining. However,

Table 7.1: *Comparison of Significant Mining Taxes of Selected Mineral Producing Countries*

Country	*Income tax (%)*	*Royalty (%)*	*VAT on imported equipment (%)*	*Typical import Duty (%)*	*Dividend W/holding Tax (%)*	*Interest W/Holding Tax (%)*	*Gov't Equity Requirement (%)*
Argentina	35	Yes	21**	1	0	15.05	None
Bolivia	25+ surtax	Yes	13**	5	12.5	12.5	None
Burkina Faso	35	Yes	18	11	12.5	12.5	10
Canada (Ontario)	31.97	None	7**	0.5	25	25	None
Chile	35/42	None	18**	10**	35	4	None
China	33	Yes	None	None	None	None	None
Ghana	35	Yes	10**	Exempt	10**	10**	10
Greenland	35	None	None	None	35	None	None
Guinea***	35	7%(market value)					
Indonesia	30	Yes	10**	None	20	20	None
Ivory Coast	35	Yes	10**	0.75	12	18	10
Kazakhstan	30	Yes	20**	None	15	15	None
Mexico	35	None	15**	10**	7.6925	15**	None
Papua New Guinea (PNG)	35/252%+(4% profits levy)[3]	10**	None	17	None	30	

[3] In PNG a profit levy of 4% has been applied to compensate for the loss in government revenue from zero-rating mineral exports for VAT purposes.

Country	*Income tax (%)*	*Royalty (%)*	*VAT on imported equipment (%)*	*Typical import Duty (%)*	*Dividend W/holding Tax (%)*	*Interest W/Holding Tax (%)*	*Gov't Equity Requirement (%)*
Peru	30	None	18**	12	None	1	None
Philippines	32	5% (gold market value)	10**	3**	15	15	None
Poland	32-28	Yes	22**	9	20	20	None
South Africa	30*	Negotiable	14**	None*	12.5	None	None
Sweden	28	None	25**	0.6**	None	None	None
Tanzania	30	3% (smelter value)	None	None	10	None	None
USA	Progressive	None	None	Varies**	30**	30**	None
Uzbekistan	33**	Yes	20**	Exempt	15	15	Sometimes
W. Australia	34	Yes	10**	5**	0-30**	10	None
Zimbabwe	35	None	None	5	20	10	None

Source: (Otto, 2000) notes: *conditions or limitations apply, **credits, refunds, exemptions & other means to reduce liability may be available.

***See Mobbs (1998)

such involvement in providing for the so categorised "public goods" which is essentially the responsibility of the public sector warrants some reward/ compensation, which could be estimated and provided for instead of, continuous incentives, which are over and above the necessary cost an idea, which is shared by other experts (Financial Times, August 8-14, 2001). Furthermore, compared to Ghana and South Africa where most of the mining companies are local (Ashanti and AngloGold respectively), Tanzania does not benefit relatively much from mining due to the fact that the majority of mining companies are from outside the country; hence capital flight is high adversely affecting domestic savings.

Table 3.8 in chapter 3 shows the amount of royalties earned by the government from the sale of gold through royalty payment. Comparing the royalty rates charged by various mining countries, Tanzania could charge a higher royalty rate and still be very competitive. One way of improving the country's benefit would be to impose a 5% royalty rate like the one charged for diamonds, the amount of income the government would get is quite significant especially with the increasing production trend in the country. With the present incentive arrangements, the government is losing a sizeable amount of its "would be" earnings by charging a lower royalty rate of 3% for gold at smelter value and other generous tax exemptions (See Table 7.2 below).

Table 7.2: *Comparison of Government Royalty Revenue from Gold Under Different Rates*

Type of Mineral	*1999 Royalty at 3%*		*1999 royalty at 5%*		*Difference of Amount of Royalty Between the 2 Rates*	
	Tshs million	*US $ million*	*Tshs million*	*US$ million*	*Tshs million*	*US$ million*
Gold exports by large scale miners	942.3	1.178	1570.5	1.963	628.2	0.785
Gold exports by dealers	15.6	0.02	26	0.03	10.4	0.01
Total royalty collected	957.9	1.198	1596.5	1.993	638.6	0.795

To be able to compare relative competitiveness of countries in attracting investments however, the average effective tax rate provides a comprehensive and useful measure of competitiveness of diverse tax regimes. This measure is based on the overall profitability after considering the taxes. Taxes are major cost components, which affect profitability, and are a factor, which is not in the investor's control. The effective tax rate (ETR) is a measure expressed as a percentage of the effective cash flow, of all amounts payable by the company to the government i.e.

Table 7.3: *Comparison of Tax Incentives & Effective Tax Rates of Selected Mineral Producing Countries*

Country	*Loss Carry Forward*	*Loss Carry Back*	*Tax Holidays*	*Tax Credits*	*Tax Stabilization*	*Effective Tax rate @ 12%*
Argentina	5yrs	None	5yrs	Yes	30yrs	67.6%
Bolivia	Indefinite	None	None	None	None	98.7%
Burkina Faso	5yrs	None	Yes, during exploration	Yes	Term of contract	204.6%
Canada Ontario	7yrs	3yrs	None	None	None	106.8%
Chile	Indefinite	Yes	None	Yes	10yrs	56.2%
China	5yrs	None	5yrs	Unknown	None	136.5%
Ghana	Indefinite	None	None	None	None	89.5%
Greenland	Indefinite	5yrs	None	None	None	82.9%
Indonesia	8yrs	None	None	None	Yes	104.0%
Ivory Coast	5yrs	None	5yrs	Yes	Unknown	120.0%
Kazakhstan	7yrs	None	None	Unknown	None	89.6%
Mexico	10yrs	None	None	None	Yes	111.5%
Papua New Guinea	7years	None	None	Yes	Yes	121.2%
Peru	4yrs	None	2yrs	Yes	None	78.5%
Philippines	5yrs	None	5yrs	None	None	56.1%
Poland	5yrs	None	None	None	None	160.7%
South Africa	Indefinite	None	None	None	None	50.6%
Sweden	Indefinite	None	None	None	None	47.2%
Tanzania	Indefinite	None	None	None	None	95.1%
USA (Arizona)	20yrs	2yrs	None	None	None	Na
USA (Nevada)	20yrs	2yrs	None	None	None	78.7%
Uzbekistan	None	None	2 to 7 yrs	None	Yes	105.4%
West Australia	Indefinite	None	None	Yes	None	76.3%
Zimbabwe	Indefinite	None	None	None	None	73.9%

Source: Otto, 2000.

$$\text{Effective Tax Rate} = \frac{\text{Present Value of all amounts paid to government}}{\text{Present value of project before tax cash flow}}$$

Using a comparative study done by the Institute for Global Resources Policy and Management of the Colorado School of Mines (CSM) Golden, (Otto *et al,* 2000), Tanzania was found to be highly competitive among the twenty countries which represent a broad cross-section of developed and developing countries (see table 7.3). Employing discounted effective tax rates as a measure of competitiveness; Tanzania was ranked number 11 out of the 20 countries. This position implies that one still has room to manoeuvre in order to improve tax revenue for the government. The variations in the tax and other incentives imply that there are various ways of using these incentives depending on a number of factors, which include management and ease of implementation. In which case, Tanzania could leave the royalty rate as it is now, but adjust the other tax incentives to provide for increased benefits to the government and eventually to the people of Tanzania.

For example Table 7.4 shows that some countries do not impose royalty requirement on mining companies instead they impose other taxes, which more than compensates the zero royalty incentive. Mexico does not charge royalty but has 15% VAT against Tanzania's none, 35% Income Tax against 30% for Tanzania, 10% import duty against none for Tanzania, a limit of 10 years for loss carry forward compared to Tanzania's indefinite period etc. Given that the quality of gold reserves are comparable and sometimes surpass other mineral rich countries,[4] Tanzania should be able

[4] For example most of the gold ore grades in Tanzania at present are averaged between 50.96 million metric tons of grading 2.7 grams per metric ton (g/t) for Geita project and a high of 17.55 million metric tons (Mt) of indicated and inferred resources of grading 13.14 (g/t) for the largest deposit to date in Tanzania, the Bulyanhulu project under Kahama mining corporation. This compares favourably with Ghana's Obuasi mine which according to Coakley, (1998) is, "one of the richest gold mines in the world which used to produce gold ore grades averaging 20 grams per metric ton of gold for much of the mine's century long history, but have declined to around 7 g/t in recent years". Indicated gold resources at other mines in Ghana include 21.7 Mt grading 2.7 g/t at Bibiani, 20.9 Mt grading 1.5 g/t at Iduapriem and 3.4 Mt grading 1.5 at Ayanfuri.

Table 7.4: *Comparison of Tax Incentives of Non-Royalty Charging Mineral Producing Countries*

Country	*Royalty*	*VAT (%)*	*Income Tax (%)*	*Import Duty*	*Dividend W/Holding Tax (%)*	*Loss Carry Forward (Years)*	*Interest W/Holding Tax (%)*	*Tax Holidays*
Canada	None	7	31.97	0.5	25	7	25	None
Chile	None	18	35/42	10	35	Indefinite	4	None
Greenland	None	None	35	None	35	Indefinite	None	None
Mexico	None	15	35	10	7.7	10	15	None
Peru	None	18	30	12	None	4	1	2
Sweden	None	25	28	0.6	None	Indefinite	None	None
USA	None	None	Progressive	Varies	30	20	30	None
Zimbabwe	None	None	35	5	20	Indefinite	10	None
Tanzania	3%	None	30	None	10	Indefinite	None	None

Extracted from table 7.1

to improve her share of benefits from mining without jeopardising her competitiveness. The Bulyanhulu mining project was described by its owners, Barrick Gold Corporation of Canada (one of the worlds giants in gold mining and trade), as the 2nd richest gold find for Barrick in their Global operation and the most cost effective project in the company's portfolio which is partly financed and guaranteed by the World Bank.[5]

Fears of Cheating and Allegations on Smuggling by Investors

There is a widespread fear of cheating and smuggling in mining activities by investors in Tanzania. This suspicion is being made real by the nature of secrecy shrouding the activity. When a container full of gold ore sunk into the Indian Ocean at Tanga Harbour in 2001 (The African, September 20, 2001:10), those who followed up the saga could only speculate that may be the ore was for testing because this had normally been the explanation about such numerous complaints about such occurrences previously. Most could not believe that the ore was for export (sale) and not for testing or processing. This raises other questions like how does one determine the value to sell and therefore to base the royalty and other taxes on! Given the state of affairs, speculation range from lack of seriousness, competence and outright corruption on the part of responsible civil servants. Indeed, the explanation given has always been Tanzania lacks processing industries hence the ore has to be shipped outside. However, this defeats the whole purpose of taking bold steps to privatise the sector, whilst no efforts is made to promote local capacity within the country or the region so as to process the ore here. Tanzania can make a notable progress in terms of realisation of benefits accruing from the mining sector by promoting and therefore expanding lapidary industry in the country. A similar approach has produced positive results in other countries in the Asian continent including Thailand and Sri Lanka. Apart from creating employment for the indigenous people, lapidaries will improve the product value through processing and therefore export earnings resulting from processes minerals. Recently, higher officials in

---

[5] Proven gold reserves in 1999 were 3.6 million troy ounces, in 2000 they were 7.5 million and 2001 they stood at 10.5 million troy ounces. (Business Times, July 27-August 2nd, 2001).

the government also began to question after, they themselves being upfront with the defensive answers. Another issue, which has been ignored throughout, is that while the focus is on the ore that is shipped outside, little is said about the sand component of the shipped gold ore. No one knows if it is costed at all. There is a tendency among policy makers to disregard the sand component as if it is not part of the resources of Tanzania. This is unfortunate because this rich national resource is shipped out at no apparent cost recovery measure.

The issue of shipment of mineral ore for processing outside Tanzania took a new and surprising turn when the Prime Minister, who was one of the staunchest defenders of the decision to allow companies to send ore outside for processing, began to question and express concern about the whole issue. However, Prime Minister did not go further than raising his concern and the ore is still shipped outside, and Tanzania continues to loose an opportunity to add value to the minerals locally.

Fears of gemstones and other minerals being illegally smuggled continue to be aired by stakeholders of the mining sector. Along the Tanzania-Kenya border, gemstones are being smuggled by foot or bicycle through Mkomazi. The existence of airstrips in mining areas such as Geita, Kalalani among others, coupled with reports of low flying aircraft landing and taking off at night in these areas which have inadequate staffing for migration, customs and security personnel raises further these fears. What is of more concern is the nature of working arrangement that exists between the various authorities responsible for border control services such as customs, immigration, security and civil aviation. The Tanzania Civil Aviation Authority (TCAA) has the mandate to monitor in-coming and out-going aircraft across our borders, while the customs department and immigration departments are charged with the responsibilities of checking imports and exports also emigrants and immigrants respectively. The fact that these aircrafts could land in airstrips, which do not have the above control services, leaves a loophole for low flying aircraft to move in and out without their occupants and cargo being properly screened and verified.

In the Panama News, volume 8, number 11 of June 16-29, 2002, George Soros called for a mandatory disclosure of payments to and transactions with Governments by multinational natural resource companies, their subsidiaries and business partners in order to curb misappropriation and abuse of revenues by Governments accrued from mining, gas and oil companies simply because these companies and other stakeholders can not control how governments spend taxes, royalties and fees. Mr Soros argues that, the state and other institutions that manage these resources are often, in practice, unaccountable to the parliaments and ordinary citizens of their countries. Further more he adds,

> "...Revenues from resource extraction are disclosed neither by the governments nor the companies involved. This lack of accountability facilitates embezzlement, corruption and revenue misappropriation..."

There is an emerging consensus within the international community in favour of corporate social responsibility and increased transparency, as is evidenced by for example, the recent European Parliament Resolution, guidelines adopted by the OECD, the UN Secretary General's Global Compact, the Global Reporting Initiative, and the International Chamber of Commerce.

Forty two local and International NGOs in more than 19 counties have endorsed a proposal for publicly traded resource companies be required by regulators to disclose aggregate information about taxes, royalties, fees and other transactions with governments and/or public sector entities for the products of every country in which these companies operate.[6] They

---

[6] These NGOs include African Oil Policy Initiative Group, African Centre for Constitutional Development (Nigeria), Agencia Ecumenica para o Desenvolvimento Social em Angola, Amnesty International, UK, Botswana Council of NGOs, Bureau of Human Rights and Rule of Law (Kyrgyzstan), CEE Bank watch Network (Czech Republic), Centre for Economic and Social Rights (USA), Christian Aid, Concern for Development Initiatives in Africa (Tanzania)

Environmental Defence (US), Fatal Transactions (Belgium), Forest Peoples Programme (UK), Friends of the Earth, UK & US, Human Rights Watch, Intermon Oxfam, Spain, Lawyers' Environmental Action Team, Tanzania, Medico International (Germany), Mineral Policy Institute (Australia), Novib (Netherlands), OIKOS (Angola), Operation Young Vote (Zambia), Oxfam America, Oxfam (GB), Partnership Africa Canada, Public Service Accountability Monitor (South Africa), Save the Children (UK), Tanzania Association of NGOs, (Tanzania), Transparency International, United Nations Association, Wassa Association of Communities Affected by Mining (Ghana) etc. (Soros, 2002).

call upon the leadership of governments and multilateral institutions to help empower civil society to hold governments accountable. They assert that, mandating disclosure of tax, fee, royalty and revenue sharing payments would contribute to the development of transparency mechanisms to track revenues, thereby helping to ensure that these revenues are directed to proper investment in growth and poverty reduction.

Employment and Remuneration

Large-scale mining has generated employment for a significant number of people but not as many as those who were employed by small-scale mining activities. Estimates show that small-scale mining employed between 500,000 to 1,5000,000 people in the years prior to 1998. With the advent of large scale-mining companies, most small-scale mining areas have been taken over by large-scale mining. Given the nature of capital intensity of large-scale mining, less people are being employed by them relative to small-scale mining. It is important to note that, large-scale mining has the technological advantage over small-scale mining in terms of exploitation of minerals lying deep underground. One emerging negative aspect of large-scale mining is in terms of remuneration particularly the problem arises where differential treatment is accorded to local and foreign employees! This again is similar to the UN system on remuneration of local and foreign staff. Anecdotal evidence shows that local employees in the big mining companies are paid substantially less compared to their foreign counterparts with similar qualifications. A foreign expert can be paid 3 to 5 times his local counterpart! This is surely a recipe for social unrest and has resulted in unstable work relations in these mines. Strange as it may seem, a foreign junior staff in terms of qualifications and experience and even organisational position gets paid a higher salary than his superior local employee. Apart from being foreign and hence requiring consideration for market conditions at home, there is no justification for this wide disparity and unfair segregation approach to remuneration and especially the humiliation of being underpaid compared to junior staff. It should be noted that the first consideration for employment is professional qualification then followed by experience. Thus such unjustly differences should not be tolerated because it undermines the dignity and respect of sovereign state and citizenry and should therefore be removed as a way to address the equitable distribution ofbenefits. The argument that foreign experts deserve higher pay because of the living standards of where they come from, is tantamount to upholding the status quo of income disparities where developing countries subsidise the living standards of the rich while they continue to languish in poverty.

Benefit Sharing with Respect to Communities in Mining Areas

The issue of the distribution of mining benefits to communities in resource rich areas has been given attention in many countries particularly in trying to accommodate them within the sharing matrix. Since in most countries mineral resources are in the custody of the state for its citizens, governments therefore play a key role in the development of mining projects. In Papua New Guinea, mining leases must be negotiated with national and provisional governments as well as with customary landowners. In the Philippines, the basic government share of mining benefits are divided into 3 major categories of; payment to national government, payment to local governments and payments to other Filipinos. The later category includes royalties to indigenous cultural communities.

In Tanzania, mining investment projects are negotiated between the central government and the mining companies. The mineral policy of Tanzania states that, "the government's objective is to ensure that there is greater involvement and participation of local communities *in the implementation of mining projects.*" The objective of this approach as the policy explains, is to facilitate the mining projects' sustainability. The policy further outlines the strategies for strengthening community participation and involvement in mining to include facilitating the establishment of strategic partnerships between mining companies and the local community. Nowhere in the policy is mentioned of royalty payment to local communities, let alone district councils where the mineral resources are to be found![7]

---

[7] The local government reforms puts more responsibility on local governments to facilitate development of their areas of jurisdiction, however, most major resources found in their localities are still controlled centrally without the local governments having a say to their utilisation and benefit distribution.

Since the advent of foreign investors in Tanzania, a lot of complaints from communities of small-scale miners and other people have been heard. Here the problem has been that of unfair eviction of small-scale miners from their sometimes lawfully acquired mining 'claims' by large-scale miners. Other complaints have been associated with the distribution of mining benefits to communities around the mining areas. In mid 2001, 14 members of parliament from the Lake Victoria region lodged their demand on mineral royalties for their constituencies, which are rich in gold and diamonds (Mtanzania, number 1902, July 17,2001). They further pressurised the government to revise the mining act so as to benefit local people in mining areas because the present one does not! In addition, they requested that the arrogance of investors should also be checked. Such complaints and demands are a manifestation of a missing link between the government and its people. Customary rights of indigenous people need to be recognised in order for any development, which takes place in their area to have meaning to them. Benefits to the communities should not be left to the discretion of the investor.

Mining Conflicts

The issue of conflicts between foreign investors and local miners is illustrated by conflicts in the year 2001 that were widely reported in Mererani between a South African Gemstone investor known as AFGEM and the small-scale miners on the one hand and the village authorities on the other and similarly between the Kakola community in Bulyanhulu and Kahama Mining Corporation.

The Mererani conflict began when the government granted mineral rights to AFGEM in Block B in Mererani, adjacent toBlock C, which was occupied by many small-scale miners. Cases on encroachment by small-scale miners led to AFGEM guards shooting and unleashing dogs at suspected "intruders." Since local miners had very small plots in Block C, they felt that it was unfair to allocate a whole block to a foreign company. On the other hand, the foreign company felt that it had the right to defend its "lawful" plot against intruders.

A second issue that escalated the conflict was the matter of AFGEM wanting to brand tanzanite. There was a feeling among the small-scale miners and other Tanzanians that by being allowed to brand the unique gemstone, AFGEM would gain monopoly in the market and that Tanzania would lose out. There were thus criticisms against allowing AFGEM the right to exclusively brand a unique Tanzanian product.

The third area of conflict arose when AFGEM questioned the rationale of being requested to contribute local village development activities such as provision of water and other village developments. While AFGEM argued that it was not part of the contract with the government, local village leaders had the notion that big business had a responsibility to assist in the development of local communities.

In all three types of conflicts the government intervened often by not addressing the core problem, namely unequal access to national resources and seemingly favouritism extended to foreign companies. Elsewhere, government intervention has taken a heavy hand by condemning those who are causing conflicts. Many people are still question why is the government taking such a protective approach to foreign companies while it pays little attention to the plight of local communities? Local communities are constantly asking for whose benefits are the minerals of Tanzania?

On a different scale (Business Times, August 24-30,2001), villagers from Kakola Division at Bulyanhulu threatened to sue the Kahama Mining Corporation limited at world court. The community claimed that the mining firm illegally occupied their land. It also wanted compensation for unlawful eviction, and demolition of the community homes and pits by the firm.

Thus the trend in conflicts observed during1998 continued to escalate, forcing the government to intervene. However, it can be said that the conflict resolution has not been able to satisfy the communities; in most cases it has tended to favour the investors who at the end of the day are awarded the right to mine.

Compensation Issues

Despite claims that mining business is generating significant income to the country and the people, there are still concerns regarding compensation. The Mineral Policy recognises compensation but in practice, this has been haphazard and left to individuals to find their way out. This has been one of the thorniest issues from the local people and other stakeholders, and one that will remain a critical problem unless more transparent and participatory process of mineral acquisition is in place.

For example, small miners in the areas that have been taken by the large-scale mining companies have had problems regarding compensation. Too often, one of the claim owner whose claim is mined by small miners may decide to surrender his claim to the larger scale mining company, often without telling the small miners or deciding about their fate. Transactions would be conducted and "ownership" would change hands, at which point, small miners will come to learn about their fate. The new claim owners are often not obliged to pay compensation to small-scale miners. Some small miners have been displaced and dispossessed of their means of livelihoods. This was witnessed in the Lake Victoria area and is also happening around Mererani.

Yet another issue is that due to inadequate knowledge about the value of their claims, local claim owners often end up selling their claims at throwaway prices. This problem is compounded by the lack of support from the Government or any other institution, including NGOs to help the local claim miners do proper evaluate their claims before selling or even advising them to enter into joint venture with the large scale miners instead of outright sell and dispossession. Thus in this situation, local claim owners are not properly compensated and often they end up poorer than they were before.

The issue of compensation has also not been resolved in the Bulyanhulu mining area where contentious issues of people being buried alive and counter argument that this has not happened are still hotly debated. Although legal restrictions prohibits us to debate this issue as the case is still running in the courts of Tanzania, suffice to say that, one of the issues raised by those who claim that some people were buried alive is the lack of fair compensation is the loss of life. It is important to stress that fair compensation if the right of those who have rightfully acquire property, which is either changing hands and or acquired for other uses.

Accidents

Accidents have continued to plague small-scale mining activities, especially in Mererani. In April 2001, there was another big loss of life in Mererani due to underground blasting. In response, the government appointed a commission of enquiry to investigate the cause of the accident and to suggest solutions. The commission presented its report to the government in June 2002 with wide-ranging recommendations on how to avoid such accidents in future. The recommendations have led to the temporary closure of mines in Mererani unless and until protective measures were in place. As a result of the recommendations, the government is stepping up mining inspection and enforcement of regulations. The temporary ban however was not well received by the women of the mineral workers who demonstrated and called for the government to un-ban the mining activity so as to allow their husbands get money for subsistence purposes. However, the public feeling on mining accidents is one of helplessness, as shown by one of the local papers that reported, "We are tired of hearing about accidents in Mererani." The occurrence of accidents at Mererani further raises questions as to whether the government is serious with security issues at the mining areas. Why has the government not addressed the issue of security comprehensively?

8

CONCLUSIONS AND RECOMMENDATIONS

This study analysed the relationship between the macroeconomic policy reforms and the economic, social and environmental aspects of the mining activity. Macroeconomic reforms in the mining sector have generated considerable benefits to the local communities, government and mining companies or groups, but more revenue could have been obtained had adequate efforts been made to create efficient markets. However, policy reforms in this sector have stimulated the growth of auxiliary activities, opened up new mining areas and increased employment opportunities.

Despite the notable economic benefits accruing from this sector, there are still a number of critical factors which require immediate attention if it is to contribute more. As observed earlier, some of the successes in the sector stem from putting in place correct policies. In addition, some of the negative consequences are to a greater extent the result of policy implementation failure or the failure of the policy to take those issues into account.

From the preceding analysis of the findings the following conclusions can be drawn.

- The significant positive influence of liberalisation of the mining sector can be seen in the growth of investment in the sector as well as the sectoral GDP share. It is indicated that there are more new approved investment licences and employment has increased. In addition, there is an expansion of economic activities in small scale and large-scale mining areas. Auxiliary activities such as transport and trade have expanded.

This confirms our hypothesis that "*liberalisation of the mining sector has stimulated mining activities and hence increased the sectoral contribution to the national output.*"

- Price regulating regimes (particularly under-pricing of minerals) and inadequate markets have cost the country heavily because this has encouraged illegal mineral exports. Inefficient marketing systems in Tanzania have tended to frustrate some mineral dealers, who then decide to smuggle minerals to other countries, to benefit from their markets (that are reliable with attractive prices). In addition, due to inadequate marketing channels the government has not been able to capture all the revenue from this sector. Prior to 1990-94, for instance, Tanzania was losing an estimated 3 tonnes of gold annually through illegal trade. This loss was reduced in 1990-94, when the government allowed some national banks to purchase gold from the dealers. Following government policy to withdraw from direct business, the banks stopped purchasing gold and, between 1995-98, the country was losing an estimated 2.8 tonnes of gold every year through smuggling. A meagre 100 kg of gold was officially recorded as annual gold recovered during the period 1986-89, instead of the actual annual production estimated to stand at 3,235 kg. Between 1990 and 1994 the amount reported was almost equal to the actual production. Afterwards only 347 kg are reported instead of 3,235 kg. This is in line with the hypothesis that "*that lack of reliable marketing channels has resulted in illegal exports, and therefore what is captured by official statistics is far below the actual recovery of minerals.*"

- Apart from showing that macroeconomic reforms have stimulated the expansion of activities in the mining sector, the study proves further that the liberalised policy regime has simplified the process of granting mineral rights. Both small-scale and large-scale miners' licence application procedures and the mineral dealers' and lapidary licensing procedures have made it much easier to obtain licences than used to be the case prior to liberalisation.

- In addition, macroeconomic policy reforms in the mining sector have stimulated movement to, and demographic changes in, the mining areas. Between 15 - 50 % of people in mining areas moved into them after 1985. The 1980s therefore witnessed an influx of small-scale miners into the mining areas in the country, with the result that they have expanded in size and population, creating markets for food as well as manufactured products. The new market levels of demand and price signals have tended to exert more pressure on natural resources and the environment, charcoal and firewood traders, as well as the construction industry being quick to exploit the surrounding forest resources. This expansion has had negative implications on the natural resource base which supports the hypothesis that *"movement of people into mining areas and the increasing number of activities will result in environmental degradation in the mining areas."*

- The analysis on social issues support the hypothesis that, *"macroeconomic policy reforms on mining have not considered the negative social and health impacts of developing this sector."* On a positive note, this study shows that social integration has been achieved in terms of people from diverse cultures and backgrounds learning to live together. Indeed, places like Mererani, Kalalani and Rwamgasa are "culturally cosmopolitan" in terms of the diversity of ethnic composition. However, and this is a contradiction (but an actual social process), there is an equal number of negative impacts on the social sphere in the mining areas. These social problems include the spread of diseases (due to poor provision of social services - water, health and congestion of people). Other social problems include incidences of rape, fighting, child labour, loss of life due to accidents and drug abuse (beers, marijuana and cocaine).

- The results of this study further support the hypothesis that, *"inadequate involvement of local authorities in matters related to mining activities in their areas have tended to exacerbate disputes."* There are numerous conflicts in the mining areas, which involve mining companies and villagers; conflicts between small-scale miners and the villagers and between small-scale miners and small-scale farmers. These conflicts are exacerbated by the fact that the local authorities and communities are not fully involved in the decision-making process on matters pertaining to the mining activities. Most of the decisions on mining are taken at higher levels (zonal or ministerial level).

- Although temporal and quantitative data on income among artisanal miners was not obtained, anecdotal evidence from the study supports the hypothesis that, *"lack of adequate information and inability to participate in the market has caused large disparities in income among artisanal miners."* Empirical data shows that markets in the mineral sector have the tendency to discriminate against small-scale miners and particularly mining workers. This is a weakness in the structure of the marketing system and an indication of weak enforcement of marketing regulations. As a result, mineral brokers (both legal and illegal) benefit more at the expense of the toiling miners.

Benefit Sharing

In addition, this study shows that income generated in the mining activities is generally not ploughed back to develop infrastructure in the respective areas, particularly those which are used by small-scale miners. The assumption that mining sites are temporary areas means that they do not attract substantial investment in social and economic infrastructure, although large-scale mining companies are taking measures to make such improvement. For example, Ashanti Goldfields has contributed towards the rehabilitation of the District Hospital in Geita. In Mugusu, Pangea is assisting in building classrooms for the local school. However, in Umba, relations between the large mining company and the neighbouring community need to be improved.

Poverty in many areas is the root cause for movements of a large number of people to the mining sites. The emergence of unplanned settlements and the inability to cope with increasing demand for social services indicate the lack of a holistic planning process, which takes on board the main pillars of sustainable development.

Tax Holidays as an Investment Incentive

Unlike tax relief or investment relief, the tax holiday frees an investor completely from the tax burden for a specified period. It is a tax amnesty as Davis, *et al* (1997) put it. The mineral policy of Tanzania and the Tanzania Investment Centre stresses government resolve to formulate and implement a mining taxation regime which will attract investors. At the same time, the central objective in the country's fiscal policy formulation is to balance the national interest with those of the investors by ensuring that the mining tax regime is equitable, stable and predictable, undistortioned and competitive both locally and internationally. These are ambitious objectives, which cannot be attained easily without jeopardising the interests of any of the stakeholders.

In comparison with other countries, this is overly generous, which puts the country at a disadvantage. Investors may be attracted by these regimes and invest heavily in the first five years but close their business after that time. In other words, the so-called attractive investment environment might benefit the investors and leave the country with nothing but empty pits and huge environmental problems.

The existing investment incentive package does not appear to provide indications of yielding positive results in the long run. Mining is an extractive business, which can lead to serious environmental problems if not well controlled. Tax holidays deny the government the revenue badly needed for development purposes thus distorting fiscal stability. Tax holidays discriminate against existing local investors, stifling local initiatives and the capacity to invest. The gap between local and foreign investors will widen, thus frustrating the ongoing efforts to reduce poverty in the country.

Capacity to Monitor and co-ordinate Mining Activities

Tanzania has very little capacity to monitor and co-ordinate mining activities which limits the country's chances of receiving more revenue for sustainable development.

Institutional Co-ordination

The study results indicate that there are numerous institutional weaknesses in the management of the mining sector. For example, the lack of involvement of local communities is a serious institutional problem because there is no clear mechanism in the set-up which allows local communities to be involved. In addition, local government (district councils and villages) is not involved in the planning and decision-making process in mining development.

The field study further shows that the interests of private investors in the mineral sector override the public interest.

The Importance of a Strong Lapidary Industry

There are few lapidaries in Tanzania. This state of affairs denies the country substantial income because Tanzania fails to add value to its minerals. There is an urgent need to create an attractive environment for the establishment of a strong lapidary industry in the country which will add more value to the minerals, provide markets to minerals dealers and, in particular, small-scale miners, increase employment and enhance development of technology in the country.

The few mineral dealers who have established their own lapidaries in Tanzania employ locals as well as experts from other countries with more experience in mineral processing, such as Sri Lanka and Thailand. Most of the machines and therefore technology are imported. The export prices of the cut stones are significantly higher than the prices of uncut stones. Both male and females are employed in the few lapidaries established in Tanzania.

Importance of Stream-lining Marketing Arrangements

Marketing of minerals is one of the major infrastructural problems in Tanzania. It is probably the major cause of smuggling and illegal trading of minerals. The joint South Africa-Tanzania project of Meremeta has changed the situation by creating a reliable market for gold in Tanzania. This study was unable to obtain marketing margins at each level of the marketing chain but it is assumed that a bigger share goes to mineral dealers at the expense of producers. A market information system is lacking, but Meremeta, seem to be in touch with world markets and set prices according to ruling world market prices.

Mining and Environmental Consequences

Environmental consequences of expanded mining activities, especially by artisanal miners, are becoming more evident. Lack of technology and skills among such mine workers lead to environmentally-unfriendly mining methods, e.g. digging of pits and abandoning them without refilling them. There is also loss of vegetation in surrounding areas, due to increased consumption of wood products by large numbers of people in mining settlements. In addition, the unsafe use of chemicals such as mercury is contaminating water sources, threatening fauna, flora and human health. Specifically, this study has found that mining activities have negative effects on:

The Landscape

One environmental problem, which is obvious in the study areas, are the pits, which are abandoned as soon as the minerals are exhausted and the miners move to more lucrative mining sites, without refilling these pits. It is claimed that the high costs of refilling and low enforcement of regulations are the main reasons behind this behaviour.

Mining and Processing Activities in Rivers

Mining activities have been observed to take place in riverbeds in Kigwasi (Tanga) and in Mugusu (Geita). This practice has led to weakening of embankments, collapsing of riverbanks, thus exacerbating erosion and siltation downstream. In addition, mineral-processing activities near the rivers (e.g. sluicing and washing) pollutes the rivers, thus causing further health risks to riparian communities. The key issue of concern in the case of gold-mining is the subsequent accumulation in sediments of mercury and its bio-magnification in the food chain within the aquatic ecosystem. This problem raises the issue of co-ordination between the relevant authorities, in this case the Ministry of Energy and Minerals, Ministry of Water, Ministry of Local Government and the Village/Local communities within mining areas.

Mining in Protected Areas

Another issue of concern is the impact of mining on sensitive habitats. Of particular concern here is mining in protected areas such as forest reserves, national parks and water catchment areas. In this study, it was found that gold-mining in Mugusu is located in a forest reserve. Mining in Kigwasi and Umba gemstone areas is taking place illegally in Mkomazi Game Reserve. However, the problem is not mining in protected areas *per se* but rather the lack of, or weak, institutional co-ordination and inadequate capacity to monitor and enforce regulations. Regulations are clear as to what should be done, should mining take place in a protected area. What happens here is that such regulations are not adhered to.

Environmental Hazards and Diseases

Another area of concern is the prevalence of environmental hazards and diseases to which people in mining communities are exposed. In Mererani, there are many cases of exposure to graphite powder and tailings that may cause lung diseases. In addition, miners and other people in mining sites are exposed to occupational hazards such as accidents caused by underground blasting, falling rocks, inadequate ventilation. The issue here is that enforcement of rules and regulations with regard to safety and environmental standards in mining areas is weak.

Policy Implications and Recommendations

Macroeconomic policy reforms in the mining industry have generated benefits in economic spheres and partly in the social sphere (in terms of creating employment, changing peoples' attitude towards life, helping create economic and social infrastructure in some of the remote areas).

Policy Failure

Benefit Sharing

However, the policy reforms have had little positive impact in improving the livelihood of many small-scale miners. This is a general policy failure that will require attention as it is not clear how small-scale miners can be assisted in improving their businesses and livelihoods. It is also recommended that the existing policy framework should be recast in order to improve marketing, institutional co-ordination, enforcement of regulations, capacity-building (in management processing, marketing strategies and environmental care). Other areas that need attention include improvement in monitoring and tax collection. In addition, the policy should define how benefits can be shared between government, claim holder and local communities.

Information Sharing

The mineral policy is not known and accessible by many stakeholders. This is a failure in dissemination and distribution. For the policy to be fully contents, it must be known and available to all stakeholders to raise awareness of the implementers. Government officials know most policies in this country, but the majority of stakeholders who are the end-users of these policies are not informed. In addition, most of the policies are written in a language that most people cannot understand.

IMPLEMENTATION FAILURE

Markets

Lack of markets and information about markets will continue to affect the mining sector, especially small-scale miners. Apart from encouraging illegal trade, the current market framework mainly benefits non-miners such as brokers. Current weaknesses in markets also tend to widen income inequality in the mining sector. It is recommended that the government, in collaboration with others, work out modalities that would encourage more private involvement in the marketing of minerals. In the short run the authorities could be encouraging private banks to buy minerals from the dealers.

Conflicts Over Mining Rights

Inadequate involvement of stakeholders partly accounts for some conflicts over mining rights, which may cause unnecessary social tension and loss of revenue. While the policy is supportive of small-scale mining, it has not defined the mechanism as to how small-scale miners can be involved in decision-making processes. Often, mine workers are the losers. It is recommended that local governments be involved in allocation of mining plots. In addition, it is also recommended that capacity-building in education and awareness-raising should be undertaken. Education about mining and environmental issues, marketing and technology transfer should be given to mine workers, instead of claim holders who rarely work in the mines.[27]

In addition, it is recommended that legal knowledge should be provided for the miners. This will help raise their capacity to understand their rights and obligations, and hence reduce clashes.

Enforcement of Regulations

Whilst the law is clear that claim holders (small-scale miners or large-scale mining companies) should fill their pits as mining comes to an end, this regulation is hardly ever enforced. This is attributed partly to the implementation of the civil service reform programme, which has resulted in a sharp decline of manpower in the law enforcement organs. Broadly speaking, this is also a failure in the policy reform for not taking into account the implications of civil service reforms as the government expands the mining sector. In order to address this problem, it is recommended that employment should be unfrozen.

This should then be followed by capacity-building in the mineral sector in order to raise the ability of law enforcement organs of the government to cope with the expansion in the sector.

It is further recommended that those mining regulations, such as size of pits, security measures and safety measures, are adequately addressed.

Institutional Co-ordination

Inadequate institutional co-ordination is a major problem that requires immediate attention. For example, the entire process of obtaining mineral

[27] The normal practice is to invite claim holders to training seminars, leaving out the actual people who work in the mine pits. This is an identification problem of whom to target for importing particular skills and knowledge and therefore it should be considered carefully.

rights is shrouded in confusion and bureaucracy. In addition, there is a profound conflict of mandates and responsibilities between the Ministry of Lands and Human Settlement, Ministry of Tourism and Natural Resources and the Ministry of Minerals and Energy. The clashes between surface and underground rights are a result of inadequate consultation and ministerial co-ordination and are evident when there are both surface and underground rights in the same area. It is recommended that the government should address and clarify institutional responsibilities in order to reduce overlap and conflicts.

Infrastructure Development

Another major problem is poor infrastructure for the mining industry. It is recommended that the economic infrastructure should be improved in order to open up mining areas and create more economic linkages. While some large-scale mining companies are willing to assist in improvement of infrastructure, it is recommended that economic incentives be considered to attract more mining companies to assist in improvement of infrastructure. Thus the economic provision for infrastructure such as roads, communication and energy should be addressed.

Related to this is the suggestion to equip the regional mining offices with adequate facilities and well-trained manpower to improve their monitoring capacity and overall administration of the mining activities in respective regions.

EIA and Strategic Environment Assessment

The mining policy stresses the need for EIA in every new development. This is a positive development but enforcement is still weak, especially in ensuring EIA recommendations are followed adequately. In addition, it is recommended that the policy be subjected to a strategic environmental assessment. Further, it is recommended that all stakeholders should participate in preparation of the sectoral Master Plan, which to-date the mineral sector does not have.

Environmental Committees at district level must be empowered by law to take legal action against individuals and/or companies contravening environmental regulations. This would however require specific capacity-building for the committees to be able to handle such matters.

Registration of Villages Adjacent to Mining Areas

In order to reduce land-use disputes, the government should ensure that villages that are close to the mining areas are registered and recognised. This will guarantee villagers' rights should mining companies or owners wish to extend their activities beyond existing boundaries.

Compensation

A major complaint from most of the stakeholders has been inadequate compensation for the land they have lost to mining companies. While this is a complex issue that requires a careful approach, it is recommended that claim holders or companies that wish to invest in mining should not be left to negotiate alone with the villagers about properties such as land they want to acquire for mining. Such discussions should involve lawyers, village government and other family members so as to reduce the chances for conflicts and complaints.

BIBLIOGRAPHY

Ackello-Ogutu C. and Echessah P. N. (1998): "Unrecorded Cross-Border Trade Between Tanzania and her Neighbours: Implications for Food Security; Technical Paper Number 89, USAID Dar es Salaam.

Bagachwa, M.S.D., Fanuel C. Shechambo, Hussein Sosovele, Kassim A. Kulinwa, Alexis A. Naho with Elizabeth Cromwell, (1995): *Structural Adjustment and Sustainable Development in Tanzania*. Dar es Salaam: Dar es Salaam University Press.

BOT (Bank of Tanzania) (1998a): *Economic Bulletin for the Quarter Ended 31 March* Vol. XXVIII No. 1.

BOT (Bank of Tanzania) (2000): *Monthly Economic Review*, February, Dar es Salaam, Tanzania.

BOT (Bank of Tanzania) (1998b): *Monthly Economic Review*, December, Dar es Salaam.

BOT (Bank of Tanzania) (2001): Economic Buleltin for the Quarter Ended 31 December, Vol. XXXII, No. 4.

BOT (Bank of Tanzania) (2002): Economic Bulletin for the Quarter Ended 31 March, Vol. XXXIII, No. 1.

Business Times, (1998): "The Government Works to restructure Taxation System," Tanzania's Independent Economic and Financial Weekly Newspaper, December 11, pp. 1, Dar es Salaam Tanzania.

Business Times (1999): "World's Top 10 Gold Producers Listed." A weekly Paper, Friday, May 21st pp. 9

Business Times (1999a): "Gold Protest Set to Grow." A Weekly Newspaper, Friday 16th of July, Dar es Salaam

Business Times, August 31- September 6, 2001 p.4 (*The politics of gold: How will Tanzania ever benefit from it?*)

Business Times, August 24- 30, 2001 (*Kahama mining saga; now villagers threaten to sue at world court*)

Business Times, July 27-August 2, 2001 p.4 (*Review mining policy-renegotiate mining contracts*)

Chachage, C.S.L. (1995): "The Meek shall Inherit the Earth but not the Mining Rights," in Gibbon P. (ed) Liberalised Development in Tanzania: *Studies in Accumulation Processes and Local Institutions.*" The Nordic Institute for African Studies, Uppsala, Sweden.

Chachage, C.S.L. (1995a): "Mining and Environmental Issues Under SAPs in Tanzania: Examples From Three Case Studies" in Bagachwa, M.S.D. and F. Limbu (eds.) *Policy Reform and the Environment in Tanzania*. Dar es Salaam University Press, Dar es Salaam, Tanzania.

Coakley, G.J (1998): The Mineral Industry of Ghana URL http:/www.ghanaweb.com/GhanaHomePage/Ghana.....

Coakley, G.J (1998): The Mineral Industry of Tanzania URL http:/www.newafrica.com/mining/Tanzania_miningdev.htm

Comparative Study." Institute of Global Resources Policy and Management, Colorado School of Mines, Golden, Colorado, USA.

Daily Mail (1999): "Mining Regulations are Finally Out." The Daily Paper, Monday the 9th of August, pp. 1, Dar es Salaam.

Daily News (1998): "Sri-Lankan Miners Come to Tanzania"; Daily Paper, Friday 8th August.

Daily News (1999): "Mining Still the Life Line of SADC States." A Daily Paper, June 22 p.1.

Davis, J., Vehorn, C., MacMorran, R and Isaac, J. (1997): "Tanzania: Strengthening the Tax System"; International Monetary Fund, Fiscal Affairs Department (FAD).

De Vera, Benjamin M (2000): Sharing the Proceeds of Mineral Development: The Fiscal Regime of Philippine Mining. Paper presented at the 8th Annual Mine Safety And Environment Symposium held on November 24th at Concorde Hotel, Baguio City.

DHV Consultants BV (1998): "Small-Scale Gold Mining in the Shinyanga, Kahama and Bukombe Districts, Tanzania." The Government of the Netherlands, Royal Netherlands Embassy, Dar es Salaam, Tanzania.

ERB (Economic Research Bureau) (1989): *Tanzanian Economic Trend*; A Quarterly Review of the Economy, Vol. 1, No. 4 January, University of Dar es Salaam.

ERB (Economic Research Bureau) (1991): *Tanzanian Economic Trend*; A Quarterly Review of the Economy, Vol. 4, No.1 April, University of Dar es Salaam.

ESRF (Economic and Social Research Foundation), (1997): "Mineral Sector Development Strategy." A Report Prepared by ESRF in Collaboration with Tan-Discovery, Dar es Salaam.

FEMATA (Federation of Miners Associations of Tanzania) (1998): Tamko Rasmi la Tatu la Shirikisho la Wachimbaji Madini Tanzania, FEMATA, Dar es Salaam.

Financial Times (1999): "The 1998 Corruption Perception Index." The Weekly Paper Wednesday June 23-29, pp. 8.

Financial Times (1999a): "Government Studies Minerals Sector Taxes to Stem Smuggling," A Weekly Paper Wednesday August 11-17, pp. 3.

Financial Times, Wednesday August 8-14th, 2001: (*Mining Act needs urgent review*)

Hackett, S.C (1998): Environmental and Natural Resources Economics: Theory, Policy, and the Sustainable Society.

Hangi, A.Y. (1996): "Environmental Impacts of Small-Scale Mining: The Case of Melelani, Kahama, Geita and Musoma," *A CEEST Research Report Series No. 1*, Dar es Salaam.

International Monetary Fund (IMF) (1990): 'International Financial Statistics; Vol. XLIII No. 1., January 1990.

International Monetary Fund (IMF) (1993): 'International Financial Statistics; August 1993.

International Monetary Fund (IMF) (1998): 'International Financial Statistics; May 1998.

International Monetary Fund (IMF) (1999): 'International Financial Statistics; June 1999.

Islam, R. (1991): "Alleviation of Rural Poverty, Growth and Macroeconomic Policies: The Asian Experience", *EDI Working Paper*, The World Bank, Washington

Jambiya G., Kulindwa K. and Sosovele H. (1997): "Poverty and the Environment: The Case of Informal Sand-Mining, Quarrying and Lime-Making Activities in Dar es Salaam, Tanzania." *Research Report No. 97.1, REPOA*, Dar es Salaam.

JET (Journalists' Environmental Association of Tanzania) (1997): "Gold rush at Kahama, Nzega, Tarime Ignores Environmental Impacts" in *JET NEWS*, April; Dar es Salaam, Tanzania.

Kaufmann, D. and O'Connell, S. (1991): "The Macroeconomics of the Unofficial Foreign Exchange Market in Tanzania", in Chibber, A. et. al. *Economic Reform in Sub-Saharan Africa*: A World Bank Symposium, Washington D.C.

Lange, G.M and R.M. Hassan (2000): Natural Resource Accounting. Policy Brief. IUCN Regional Office for Southern Africa

Lipumba, N. H. I. and Mbelle, A. V. Y. (1993). "Policy-Making for Economic Growth and Poverty Elimination in a Pluralistic Political System in Tanzania: An Overview", in Bagachwa, M.S.D. *et. al. Economic Policy under a Multiparty System in Tanzania*; Dar es Salaam: Dar es Salaam University Press, University of Dar es Salaam, Tanzania.

Milne, G.R. and Marongwe D..A. (1996): "Small Scale Gold Panning in Zimbabwe; Economic Costs and Benefits and Policy Recommendations"; Unit for Environmental Economics, Department of Economics, Gothenburg University, Sweden,.

Mnali S. R. (1997): "Physical Environmental Pollution Resulting From Artisanal Gold Mining in the Lake Victoria Gold Fields, Tanzania", A Paper Presented at the Conference on Small-Scale Mining in African Countries, 29th September to 1st October, Dar es Salaam, Tanzania.

Mobbs, Philip M. (1998): The Mineral Industry of Guinea. URL http:/www.newsafrica.com/mining/Guinea...

Mpango, P. (1997): "Tax Administration in a Changing Economic Environment: Reflections on the Tax Reform Experience in Tanzania and the Challenges Ahead." A paper Presented to Commemorate the Silver Jubilee of the Institute of Finance Management (IFM), Dar es Salaam.

Mpango, P.I.N. (1996): Evasion of Import Duties: An Economic Analysis of the Case of Tanzania; Unpublished A Ph.D. Thesis, University of Dar es Salaam, Tanzania.

Mtanzania number 1902: Tuesday, July 17,2001 p.2 (*MPs from Lake Region on Revision of Mining Act*)

Mtanzania number 1906, Saturday, July 21, 2001 p.1 (*MPs from Lake Region on Royalty*)

Mutagwaba, W., Mwaipopo-Ako, R. and Mlaki, A. (1997): "The Impact of Technology on Poverty Alleviation: The Case of Artisanal Mining in Tanzania, *Research Report No. 97.2, REPOA*, Inter Press Tanzania Ltd., Dar es Salaam.

Nanyaro, J.T. (1994): "The Mining Sector of Tanzania: Current Status and Strategies for its Sustainable Development in the 21st Century" In: Msambichaka, A.L. *et al* (Eds*) Development Challenges and Strategies for Tanzania: An Agenda for the 21st Century.* Dar es Salaam, Tanzania.

Ndulu , B. J. (1994): "Tanzania's Economic Development: Lessons from Experience and Challenges for the Future", in Msambichaka, L. A., Moshi, H. P. B. and Mtatifikolo, F. P. (eds.), *Development Challenges and Strategies for Tanzania: An Agenda for the 21st Century*; DUP, Dar es Salaam, Tanzania.

Nipashe (1997): "Madini ya Bilioni 24/- Yauzwa kwa Magendo." A Daily Newspaper, September 23rd, p.1, Dar es Salaam, Tanzania.

Nyagetera, B. and Tarimo, B. (1997): "Financial and Monetary Sector Management during and after Reform in Tanzania", in Bol, D., Luvanga, N. and Shitundu J. (eds) *Economic Management in Tanzania*, TEMA Publishers Company Ltd, Dar es Salaam, Tanzania.

Otto, J., Batarseh, M. and Cordes, J. (2000): "Global Mining Taxation"

Panos (1996) The Lure of Gold – How Golden is the Future? http:/www.panos.org.uk/briefing/gold/html

Perman R., Ma Y. and McGilvray J. (1996). *Natural Resource and Environmental Economics*, Longman, London.

Projekt-Consult [n.d.], "Programme of Integrated Environmental Management in Small-Scale Mining in Bolivia". Mimeo.

Putterman, L. (1995). "Economic Reform and Smallholder Agriculture in Tanzania: A Discussion of Recent Market Liberalisation, Road Rehabilitation and Technology Dissemination Efforts," in *World Development*; Vol. 23, pp. 311-326.

Reed, D. (ed.) (1996): *Structural Adjustment, the Environment and Sustainable Development*. London: Earthscan.

Sankhayan, P.L. (1994). *Effects of Economic Policy Reforms on Peasant Agriculture and the Management of Natural Resources in Tanzania*; in Ecology and development Paper no. 15, Agricultural University of Norway.

Shitundu, J. and Mjema, G. (1999): "Re-starting Growth in Tanzania." A Paper Submitted to REPOA, Dar es Salaam, Tanzania.

Soros, George (2002): Publish What You Pay. The Panama News Opinion, Volume 8, number 11, June 16-29.

Sunday Observer (1997); "Gold Miners Poisoning Lake Victoria With Mercury." A Weekly Paper, October 12, pp. 3.

Sunday Observer, (1998): "Tanzania's Gemstones Siphoned Away." Tanzania's Independent Weekly Newspaper, July 19, p. 1, Dar es Salaam, Tanzania.

The African, September 20, 2001p.10 (*Sunken Ore at Tanga Harbour*)

URT (1997): The Mineral Policy of Tanzania. Ministry of Energy and Minerals, Dar es Salaam

WTO (1999): Papua New Guinea: November 1999. Trade Policy Reviews. November. http:/www.wto.org/English/tratop_e/tpr_e/tp 120_e.html

Tanzania Mineral Dealers Development Association (TAMIDA) (1999): Annual Report 1999; Arusha, Tanzania.

TIG (Tanzania Investment Guide) (1998): *Investors' Guide to Tanzania* 1998 Dar es Salaam, Tanzania.

UNDP (United Nations Development Programme) (1991): *Human Development Report 1991*; Oxford University Press, New York.

UNDP (United Nations Development Programme) (1997): *Human Development Report 1997*; Oxford University Press, New York.

UNEP (United Nations Environment Programme) (1997): "Mining and Sustainable Development" in Industry and Environment, Vol. 20 No. 4, October-December.

URT (United Republic of Tanzania), (1997a): "Environmental Management Framework for the Mineral Sector," Consultative Draft Report submitted to the Ministry of Energy and Minerals, Dar es Salaam, Tanzania.

URT (United Republic of Tanzania) (2002): Economic Survey 2001; Planning and Privatization, Dar es Salaam, Tanzania.

URT (United Republic of Tanzania) (1999): *Economic Survey 1998*; Planning Commission, Dar es Salaam, Tanzania.

URT(United Republic of Tanzania) (1998) *Economic Survey 1997*, Planning Commission, Dar es Salaam, Tanzania.

URT (United Republic of Tanzania) (1996): "The Rolling Plan and Forward Budget for Tanzania for the Period 1996/97-1998/99." Planning Commission, Dar es Salaam, Tanzania.

URT (United Republic of Tanzania) (1994a): *Economic Survey 1993*, The Planning Commission, Dar es Salaam, Tanzania.

URT (United Republic of Tanzania) (1994b): "Tanzania. National Environmental Action Plan (NEAP): A First Step," Dar es Salaam: Ministry of Tourism, Natural Resources and Environment, Tanzania.

URT (United Republic of Tanzania) (1996): Econo*mic Survey 1995*; The Planning Commission, Dar es Salaam, Tanzania.

URT (United Republic of Tanzania) (1997a): "The Mineral Policy of Tanzania; Ministry of Energy and Minerals," Dar es Salaam, Tanzania.

URT (United Republic of Tanzania) (1997b): "Speech by the Minister for Energy and Minerals, Hon. Dr. Abdallah Omari Kigoda, (MP), Introducing to the National Assembly the Ministry's Estimates of Public Revenue and Expenditure for the Financial year 1997/98," Ministry of Energy and Minerals, Dar es Salaam, Tanzania.

URT (United Republic of Tanzania) (1997c): "National Environmental Policy," Vice President's Office, Dar es Salaam, Tanzania

URT (United Republic of Tanzania) (1997d): *Economic Survey 1996*; The Planning Commission, Dar es Salaam, Tanzania.

URT (United Republic of Tanzania) (1998a): *Economic Survey* 1997; The Planning Commission, Dar Es Salaam, Tanzania.

URT (United Republic of Tanzania) (1998b): "The Mining Act, 1998; Ministry of Energy and Minerals", Dar Es Salaam, Tanzania.

URT (United Republic of Tanzania) (1998d): "Tanzania: Opportunities for Mineral Resource Development": Ministry of Energy and Minerals, Dar-Es-Salaam, Tanzania.

URT (United Republic of Tanzania) (1998e): "Taarifa ya Maendeleo ya Mkoa wa Arusha kwa Rais wa Jamhuri ya Muungano wa Tanzania," Ofisi ya Mkuu wa Mkoa wa Arusha.

URT (United Republic of Tanzania) (1998): *National Poverty Eradication Strategy (NPES)*; Vice President's Office, Dar es Salaam.

URT (United Republic of Tanzania) (1999): The Tanzania Development Vision 2025, Planning Commission, Dar-es-Salaam.

Wagao, J. H. (1993): "Multi-Partyism and the Tanzanian Economy: Some Challenges," in Bagachwa, M. S. D. et. al. *Economic Policy under a Multi-Party System in Tanzania*; DUP, Dar es Salaam, Tanzania.

Wangwe S. M. and Ndulu B. (1997): "Managing the Tanzanian Economy in Transition to Sustained Development" In: *Economic Management in Transition to Sustained Development in Tanzania,* Bol, D, N. Luvanga and J. Shitundu (eds). Dar es Salaam: TEMA Publications, 1997 p.21-45.

WWF (World Wide Fund) (1993): "Sustainable Use of Natural Resources: Concepts, Issues and Criteria," A WWF International Position, Paper, Gland, Switzerland.

APPENDIX 1: STRUCTURE AND CHARACTERISTICS OF THE STUDY AREA

An Overview of the Study Area

As noted earlier, three study areas have been selected based on their significance in terms of type of minerals available, social and economic relevance as well as the subsequent policy significance. These are Mererani in Arusha Region, Geita in Mwanza Region and Umba in Tanga Region.

Mererani in Arusha Region

Mererani is said to be the only source of tanzanite in the world. It is located in Simanjiro District about 70 kilometers south of Arusha town. Mererani is an old village, which is currently harbouring approximately 40,000 people. The village was in existence long before the discovery of gemstones. The indigenous people are Wameru whose occupation used to be farming and livestock keeping. Tanzanite was discovered by a herdsman, one Mzee Manyanya, when he was trekking his cattle towards the grazing areas. Mererani residential area is about 7 kilometers from the claim holdings or mining pits. There are 9 suburbs in Mererani, namely Cairo, Lyamtu, Low Soweto, Songambele A, Songambele B, Songambele C, Songambele D, Zaire and Kazamoyo. The area is also rich in other gemstones and graphite.

The mining area at Mererani is divided into four blocks. Block A is under Kilimanjaro Mines Company owned by a Tanzanian. Blocks B and D are allocated to small-scale miners while Block C is owned by a South African Mining Company called Graphite Tanzania (GRAPHTAN).

Following tanzanite mining at Mererani, Arusha Region has been expanding its gemstone volume of trade for the past two decades. It is reported that the mineral sector's contribution to the regional GDP is estimated to be 2% (URT 1998e), but unofficial sources show that the figure could stand between 35% and 40%. Official figures tend to be lower than reality, reflecting the extent of unrecorded trade, i.e. illegal trade and/ or smuggling.

The stakeholders in the region consist of the zonal mines office at the apex, associations, mineral dealers, claim holders (small-scale or artisanal miners and mining companies). Important institutions at Mererani include Mererani village government, mines security committees and a range of service providers.

Zonal Mines Office, Associations and Mineral Dealers

The Zonal Mines Office in Arusha caters for the entire Arusha zone, which comprises Arusha, Kilimanjaro and Tanga Regions. The office is entrusted with overseeing the enforcement of government policy and laws on mining activities in the three regions. In addition to the zonal mines office, there are four registered associations operating in the region. These are Arusha Region Miners Association (AREMA), Tanzania Mineral Dealers Association (TAMIDA), Arusha Women Miners Association (AWOMA) and Arusha Central Selling Organisation (ACSA).

Each region in the mainland where mining is carried out has formed a Regional Miners Association (REMA). In Arusha region the association is called Arusha Region Miners Association (AREMA). AREMA was created in 1989 by small-scale miners. Its primary objective is to educate members and make them aware of the importance of expanding and improving the mining activities in line with the mining policy in the interests of small-scale miners and the entire nation. At the national level, the REMAs are co-ordinated by the Federation of Miners Associations in Tanzania (FEMATA).

Tanzania Mineral Dealers Association (TAMIDA) was also formed in 1989. It is a national forum, which covers all the regions in the country. Membership consists of licensed mineral dealers some of which are also claim holders. Mineral dealers are mainly dealing with purchases, processing and marketing of minerals and TAMIDA is meant to promote mineral trade in Tanzania as well as outside. At its establishment in 1989 TAMIDA had 11 members in Tanzania. Currently, there are 120 TAMIDA members in Tanzania out of which 79 are based in Arusha region.

Tanzania Women Miners Association (TAWOMA) is a national association, which plans to initiate the formation of Regional Women Miners Associations (REWOMAs). TAWOMA was established on August 22, 1997 with the aim of educating small-scale miners, especially women, about the mining sector in general. The initial plan is to form the associations in five regions namely, Arusha, Morogoro, Dar es Salaam, Singida and Tanga. Arusha region has already registered the Arusha Women Miners Association (AWOMA). Women are extensively participating in quarrying, sand-mining, pottery, gemstone-mining, processing limestone and marketing of minerals as brokers, among other economic activities. Thus, it was necessary that they form an association, which will co-ordinate and protect their interests.

The lucrative business in the mineral sector has attracted many youths particularly in gemstone-marketing. Most of them are unregistered brokers who have the tendency to buy unprocessed stones mainly from artisanal small-scale miners and mining workers. Brokers are said to spearhead smuggling of precious stones and they do not seem to be popular with TAMIDA as their activities are considered to be interfering with mineral dealers' operations. Some of the brokers are foreigners. In the past many brokers were arrested because most of them used to operate illegally. Subsequently, the situation has became worse as riots and unemployment were on the increase. This is what prompted the establishment of Arusha Central Selling Organisation (ACSO) whose major objective is to provide a buying and selling forum for the youths under a single master dealer licence.

Small-Scale Miners (Claim Holders), Artisanal Miners and Mining Companies

The distinction between small-scale mining and artisanal mining does not seem to matter in the discussion of mining activities in Tanzania. Chachage (1995) calls small-scale miners "artisans," "galampsey," "creusseurs" or "pailleurs," implying that the words small-scale and artisanal mining are interchangeable. Hangi (1996), who also uses the words small-scale and artisanal mining interchangeably, observes that small-scale mining in Tanzania ranges from informal subsistence mining or artisanal mining to formally organised small mines which are privately or publicly owned. Mutagwaba *et al* (1997) considers the two categories of miners as one and calls them small-scale artisanal miners. DHV Consultants BV (1998) observes that there is a gradual change from artisanal that is completely manual mini-scale mining to slightly mechanised small-scale mining. Since the distinction between the two forms of mining is not clear-cut but rather transitional, they argue, the terms are used as one. UNEP (1998) on the other hand recommends distinction between the two terms. They have suggested that, for wider adoption, small-scale mining should be regarded as legal, i.e. miners who have the right to the minerals they sell, and their activities are known to and overseen by the respective government and local authorities, whereas, artisanal mining should be referred to as illegal mining, i.e. the mining of minerals by people who have not been sanctioned by the authorities. For the purpose of this report artisanal mining does not necessarily mean illegal mining. Likewise small-scale mining is not always legal as some of them may be running with expired title deeds. Small-scale and artisanal mining are therefore going to be used interchangeably.

In July 1998, 342 claim holders were working at Mererani despite the fact that there are approximately 600 claim titles (small-scale holdings) in Arusha region. Most of the titles are located at Mererani. Each of the claim holders employs an average of 20 miners, which makes a total of 6,840 mining workers at Mererani by July, 1998. As already noted earlier, in reality, workers of the mines have no job security. For instance, while small-scale miners and mineral dealers have formed their associations which are meant to protect their interests, the employees are virtually operating in a vacuum. There is no forum or institution which is owned by the workers to safeguard and promote their own benefits. Suprisingly, TAMIDA and AREMA leaders have, since their establishment, pretended to accommodate the interests of the mining workers through their respective associations. This role does not seem to be realistic because small-scale miners and mineral dealers on the one hand and the mining workers on the other are two different and opposing interest groups. How can a small-scale miners' association, eg. AREMA, pretend to cater for the interests of the mining workers, i.e. small-scale miners' employees?

Mererani consists of both small-scale as well as large-scale miners. A number of both local and foreign companies used to operate at Mererani. They include Graphite Tanzania (Graphtan), Tanzania Gemstone Industry (TGI), Kilimanjaro Mines and Building Utilities. All the four companies are now closed despite the fact that Kilimanjaro Mines and Graphtan still own blocks A and C respectively, with their equipment and buildings remaining on site.

Mererani Village Leadership, Mines Security Committee and Service Providers

The village has a strong leadership, which comprises both government and party leaders. The leadership is fully involved in all matters pertaining to mining activities in the area. The village has initiated formation of the mines security committee following a series of disputes on encroachment among claim holders. The committees are thus entrusted to oversee all the activities around the mining pits. The main obligation is to ensure that all the parties abide by the governing rules and regulations.

Various other economic activities or businesses have been emerging within the mining community. These entrepreneurs are important service providers. They range from religious leaders of different denominations, restaurants, health centres, shops, and transportation.

Geita in Mwanza Region

Mwanza, Kagera and Mara Regions belong to one zone, which used to have a total of six regions in the past. The other three regions now make an independent zone. These are Shinyanga, Kigoma and Tabora. The headquarters of the Mwanza Zone is in Mwanza town. Gold mining in Mwanza region is mainly concentrated in Geita district. Small-scale miners are concentrated in Nyakagwe, Nyarugusu, Mugusu and Rwamgasa areas. Owing to its size and the fact that all the districts are blessed with mineral reserves, each district in Mwanza region is divided into at least one sub-zone. Geita district, for example, has four sub-zones namely, Nyakagwe, Nyarugusu, Nyamtondo and Geita Urban. Two types of minerals are particularly important in this zone. These are diamond and gold.

Associations and Mineral Dealers

Mwanza Region Miners Association (MWAREMA) was established in 1986 following the 1983 government appeal that small-scale miners should look at possibilities of establishing regional miners' associations. MWAREMA maintains leadership at both sub-zonal as well as regional level.

There are roughly 550 members of MWAREMA, of which only 400 are active. Prior to nationalisation many mining areas were under private ownership. The Dar Tardine el Umma Tanzania Limited (DTT) which was operating in Geita organised miners as sub-contrators, apportioning them areas of 400 square feet to carry out mining on condition that they sold the gold to DTT. MWAREMA was subsequently formed to supervise the sub-contractors. The association also took a few claims outside the DTT's area on behalf of its members. Since DTT's licence was revoked in 1989, MWAREMA has continued to supervise the sub-contractors, although for a long period the area remained unlicensed because DTT had filed a case against the government with the intention to reirinstate their ownership of the area. MWAREMA was basically formed to promote and protect the small-scale miners' interests. Membership is free but there is an entry fee of 5,000 TShs. MWAREMA has two wings namely, Mwanza Region Mining Company Limited (MWAMICO) and Mwanza Region Women Miners Association (MWAWOMA). Both MWAMICO and MWAWOMA are still in the infant stage. MWAMICO is meant to deal with marketing of minerals and MWAWOMA is an association which caters for women's mining activities in the region. TAMIDA is not widely known in Mwanza region.

Unlike Arusha, there are only three registered mineral dealers in Mwanza region and with the exception of one, all of them are reportedly inactive. It appears that most prospective mineral dealers are encouraged to undertake their activities illegally because for them procedures to formalise the business are still tedious and bureaucratic, i.e. one can make better profit margins by operating informally.

The only stable mineral dealer is Meremeta Limited, which started on the 7th November, 1997. This is a joint company owned by the Tanzanian government (Tanzania People's Defence Force (TPDF)) and the South African-based French company called Triennex. Most of the permanent high-ranking employees are foreigners. The total number of workers is around 100 most of whom are casual labourers. They have 9 expatriates,

1 person seconded from the Ministry of Energy and Minerals and 22 soldiers from TPDF as security guards. There are widespread allegations against Meremeta that the company discourages transparency and consequently some of the activities and transactions of the company are not known. Meremeta is extensively supporting small-scale miners by providing them with working gear on the condition that they sell gold to the company. Apparently there are allegations that Meremeta exploits the situation purposely to fulfil private objectives without caring much about what benefits are destined to the community. One of the motives is to maximise profits by offering different prices to different clients. In so doing Meremeta is engaging in price discrimination. This occurs when a given product is sold at more than one price and these price differences are not justified by cost differences. Some of the receipts issued by Meremeta reportedly do not have serial numbers. Apart from that, Meremeta's programmes have been widely acknowledged and appreciated. Unlike other mineral dealers, Meremeta is exempted from certain taxes. The company is obliged to pay 3% royalties only, apparently because it is partly owned by the government.

Meremeta's programmes to support small-scale miners have recently been interrupted following repeated clashes between them and the mining companies in Geita. These conflicts seem to benefit neither small-scale miners nor the respective village communities. The two rivals are virtually conflicting at the expense of the small-scale miners and neighbouring communities and/or village members who have long been raising the alarm and protesting to all the government leaders who have been visiting this area for the past ten years.

Small-Scale Miners, Artisanal Miners and Mining Companies

The government revoked DTT's mining licence in 1989 following an allegation that the company was trading in minerals illegally. Subsequently, DTT's prospecting rights were issued to the company's sub-contractors in 1991 and the claim titles were approved in 1994. Many other applications were made particularly because there were rumours that large-scale miners (mining companies) might replace small-scale miners in the area. Thus between July and August, 1995 many other title deeds were approved. To finance the 1995 title reallocation, the prospective claim holders contributed a total of 7.3 million Tshs.

The population at Rwamgasa and Mugusu is estimated to be 25,000 and 15,000 respectively. It is almost impossible to know the exact population size in most mining villages due to the uncontrolled movements of people. There is an influx of people into areas with good production-commonly known as the gold rush. This rush is most evident where claim titles have not been issued, in areas where licences have been revoked and in areas which are still being prospected. Illegal mining is therefore rise. This practice is particularly common in Matabe, which is located in the alluvial plain of Biharamulo Forest Reserve. Illegal miners are extremely destructive causing severe damage and many hazards due to the absence of regulations to govern their activities. Accidents and loss of life are prevalent. Last year, according to the former Matabe secretary general, there was a gold rush and more than 30 people are said to have died.[28] The miners have neither an obligation nor the urge to report the shocking incidences. They have the tendency to withhold sensitive information because they do not want to give a frightening impression to the authorities, which would prompt the government to remove them.

Many large-scale miners prefer to invest in gold rather than gemstones. The total number of gemstone-mining companies in Arusha zone is less than 8, whereas in Mwanza gold-mining zone, a total of 28 mining companies are prospecting and a few have almost concluded their exploration. Large-scale investors are said to be more interested in industrial minerals as opposed to gemstones for one major reason-one can establish industrial mineral reserves before the mining work starts. Gemstone-mining is gambling in the sense that investors cannot establish the reserves beforehand.

The total number of mineral rights granted as of 1999 also shows that 83% of the rights were gold while only 5% were issued against diamond, 3% against gemstone and 9% against other types of mineral.

---

[28] These are leaders, which are elected informally, and they are location specific. They are not officially recognised by either local or central government.

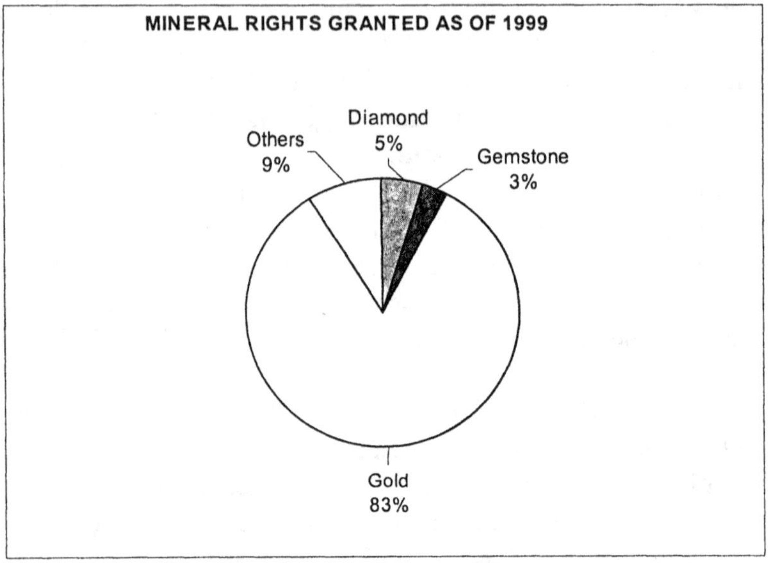

Fig. 1: *Mineral Rights Granted as of 1999*

Umba River in Tanga

Williamson Diamonds was the first company to prospect in Umba area in the 1950s. The results of this exploration revealed that there were very little reserves of diamond but massive stocks of gemstones, which did not capture the company's interest. Between 1950 and 1961 a mining company under George Poupolous, a Greek settler in Tanzania, used to operate at Umba. George became very successful in this business. Before that he was mining salt in Tanga. It was easy and cheap for George to enter into gemstone-mining because he managed to access the exploration findings from Williamson Diamonds and the fact that the indigenous people in the area, i.e. the Kamba and Maasai, were very friendly.

Many precious stones could be found on the surface during George's era and the Kamba and Maasai used to collect such stones and hand them over to George without knowing that the stones were valuable. Many people were not paid and a few of them managed to get a very low and exploitative price from George. Thousands of tons of gemstones were collected and George became very rich and famous in Europe where he is now living. It is reported that jets used to land at an airstrip in Kalalani direct from Voi to collect the precious stones and leave. George's mining licence was revoked in the early 1970s following an allegation that he was involved in illegal mineral operations. George was at once placed under arrest at Ukonga Prison after which information is missing on what transpired. Thereafter his property could not be reallocated until after 1972 when the State Mining Corporation (STAMICO) was established and entrusted to operate the mining sector in the country.

Tanga region is said to be a gemstone paradise. The most common minerals include rubies, sapphires, tourmalines and gem-garnets. Other minerals are bauxite, gypsum, gravel, salt and sand. Mica and quartz are also available.

Associations and Mineral Dealers

Many settlers joined the mining sector after retiring from the sisal industry in the early 1970s. Since 1972 all mineral dealers have been buying minerals from Tanzania Gemstone Industry (TGI), a subsidiary company of State Mining Corporation (STAMICO). To put it another way, mineral dealers were not allowed to buy directly from the small-scale miners. They were required to sell minerals to TGI who were entrusted with the control of the entire mineral trade. Like many other-state owned companies, TGI became very inefficient to the extent that it started to buy minerals from small-scale miners on credit and not paying promptly. This was an important factor which encouraged further the illegal trade of smuggling. Subsequently, the need to establish Tanga Regional Mineral Association (TAREMA) to represent the small-scale miners and protect their interests became very obvious. TAREMA has approximately 600 members, most of whom are not active. TAREMA has the obligation of ensuring that small-scale mining is developed and the quality of the products such as salt is maintained. Mobilisation of resources for its members is therefore one of the tasks of the association. Nevertheless, since most of its members are not active, TAREMA has not been very strong and effective. This is what led TAREMA to form TAREMA HOLDINGS Co. in 1997. At present TAREMA HOLDINGS has only one staff. The plans to employ directors, managers and lawyers are under way. TAREMA HOLDINGS Co. is expected to provide security for members who apply for loans. TAREMA HOLDINGS Co. will process all the loans on behalf of members after they have met all the conditions.

Apart from TAREMA, there is Tanga Women Miners Association (TAWOMA) which caters for women small-scale miners in Tanga region. The majority of women miners are involved in quarrying and salt-mining. There are only 5 mineral dealers in the region. However the regional mining office has received more than 15 new applications following the recent relaxation of licence application procedures by the government. Out of 5 mineral dealers in the region, only one of them is a member of Tanzania Mineral Dealers Association (TAMIDA) which is not widely acknowledged in the region.

Small-scale Miners, Artisanal Mining and Mining Companies

The region has approximately 600 claim titles but only 215 are operating. Small-scale miners are scattered over four different villages namely, Ng'ombeni in Daluni Ward - Muheza district, Kalalani, Kigwasi and Mwakijembe in Korogwe district. Ng'ombeni is located 73 kilometers from Tanga town and Kalalani is 15 kilometers away from Ng'ombeni.

On a number of occasions, miners, mainly from Kigwasi and Mwakijembe villages, have been encroaching on Mkomazi Game Reserve illegally for the purposes of mining. Mkomazi Game Reserve is said to be very rich in mineral reserves and many times Kigwasi and Mwakijembe residents have been requesting the authorities to allow them to apply for prospecting licences and title deeds inside the Reserve. There are cases where people from Kigwasi have illegally mined inside areas which are under the ownership of Africa-Asia Precious Stones Co. Ltd (AAPS). In 1990 two miners were shot dead by AAPS security guards when they were resisting arrest. The victims were caught mining illegally in the company's established open cast.

AAPS is the only mining company in Tanga region. The Company occupies Mikocheni area at Umba under the mining licence No. 4/90. This is a partnership between Tanzanians and Thais. AAPS are still prospecting. Down to Earth Skyline and a Russian Company, Green

Gemstones, have closed their operations due to unbearably high initial investment costs.

Gender Composition, Respondents' Level of Education and their Perception of the Natural Environment

Sample Areas

The field survey was conducted in Arusha, Mwanza and Tanga Regions. In Arusha, interviews were conducted in Arusha town and Mererani. The mineral involved is mainly tanzanite and some minor gemstones. In Mwanza interviews were conducted in Mwanza town and Geita District. In Geita, interviews were done in Geita town, Rwamgasa and Mgusu, which are gold-mining villages. In Tanga region, the study sites are located in the gemstone rich areas of Korogwe and Muheza Districts, where the study sites are Kalalani, Kigwasi, Mwakijembe and Umba Mikocheni (Korogwe District) and Daluni and Ng'ombeni (Muheza District).

Gender Composition

Out of the 88 mining labourer respondents, 80 of them (or 91%) were men and only 9 (9%) were women. This indicates the dominance of men in the mining sector. The gender composition of the sample is shown in Table 1.

Table 1: *Gender Composition of the Sampled Mining Labourers by Study Site (in %)*

Sex	*Arusha*	*Mwanza*	*Tanga*
Male	94.7	100	81.3
Female	5.3	0	18.8
Total	100	100	100

Source: Field Survey, July/August, 1998.

Education Level

As expected, the majority of the respondents are primary school-leavers, representing 75% of the respondents, compared with 11.4 % who completed secondary education and 8% with higher than secondary education. Only 5.7% of them did not attend school at all. The education levels of the respondents in the different study sites are shown in table 21 below.

Table 2: *Education Levels of the Sampled Mining Labourers (Respondents) (in %)*

Education level	*Arusha*	*Mwanza*	*Tanga*
Primary	71.1	88.2	71.9
Secondary	13.2	0	15.6
Post-Secondary	13.2	5.9	13.2
Not Schooled	2.6	5.9	9.4
Total	100	100	100

Source: Field Survey, July/August, 1998

Perception by Respondents of Environmental Problems

Using both structured and unstructured interviews, respondents were asked about environmental impacts of mining activities. These responses are used to gauge their perception about environmental impacts and solutions.

Impact on Landscape

From the structured questionnaire, there was a question on the effect of mining on the landscape. Overall, results indicate that 41% of the respondents mentioned scattered pits and holes as the most visible impact of mining. Furthermore, another 42% mentioned scattered piles of earth or rubble as another impact (aesthetics). Only 6% mentioned that pit-filling is ever practised. An insignificant proportion (8%) thought that there is no serious impact on the landscape caused by mining activities.

However, dis-aggregation of the results by area shows variations in the responses. In Arusha, no-one mentioned that pit-filling is ever done at all. No reason was given but perhaps because the mines are very deep e.g. 150-200 metres and running horizontally, miners think they ought to fill the pits. Nearly 81% mentioned scattered piles of earth, while 11% mentioned scattered pits as an environmental problem. It was noted that 8% of them thought that there was no problem with the landscape. In Mwanza 47% and 29% respectively mentioned scattered pits and piles of earth as visible impacts on the environment but close to 12% did not see any problem at all. In Tanga, only 6% of the respondents did not see a problem with the landscape arising from mining activities, but an overwhelming majority (77.4%) thought scattered pits were a problem. However, the practice of pit-filling is reported by 16% of the respondents. This result is not surprising given the fact that, at the AAPS Mine at Umba, there is a programme for refilling unproductive pits.

Awareness of the environmental implications of mining activities is gaining momentum. In Arusha, one mineral dealer pointed out that mining activities affect the environment in ways such as deforestation, changes in the landscape by creating pits and sand which creates dust.[29] According to the Natural Resources officer, miners are supposed to refill the pits, but this is hardly ever done. Lack of incentives and weak enforcement partly account for the problem.[30] The Zonal Mining Officer further emphasised the weaknesses in enforcing regulations on re-filling the pits. This is a serious problem which can lead to major problems.[31]

In Kalalani, mining of gemstones is alluvial and therefore there is not much damage to the environment. At AAPS Umba Mine, there is heavy tunneling and trenching. The policy of this company is that, as long as the mines are still productive, there is no point in filling them. However, a

[29] This is a view from the mineral dealer, Tanga, 1998

[30] Views from the Natural Resource Officer, Arusha, 1998.

[31] Zonal Mining Officer, Arusha, 1998.

number of exhausted gemstone mines have been filled up by the AAPS Umba Mine.

According to an officer in the regional planning office, a number of measures to prevent further degradation are being undertaken. One is to reserve special areas for sand-mining, unlike the earlier situation when miners used to mine wherever they wanted. Another measure is the introduction of user fees for mangroves being cut for salt-drying and other uses. It is believed that through this user fees system, the Regional Natural Resources Office gets six times more money from miners than what the Zonal Mining Office gets from the same miners.[32]

Commenting on the problems of enforcement of rules and regulations, the Chairman of Kigwasi village had the following to say: "when I got my licence back in the 1980's, the rules were quite explicit. After mining, it was mandatory to fill back the soil and we were not allowed to cut trees. These rules were followed because at that time, if you did not follow them, you were punished by paying a fine. But nowadays, things are very loose and as a result the environment is very much affected. What is happening today is that the claim title holder lives very far away from the mining area. The problem is that those who are appointed do not know the rules.[33]"

Enforcement of rules and regulations is a big problem in mining. The problem of low capacity to enforce rules and regulations for environmental protection was echoed by the Regional Planning Officer who said that, *"the capacity of the Ministry of Energy and Minerals to monitor its activities is low due to inadequate staffing, lack of transport and lack of financial resources. As a result the zonal office has failed to open mining offices in districts which are active in mining, e.g. Handeni.*[34]" However, some efforts are being made to prevent further

[32] Report from the Regional Planning Office, Tanga, 1998. However, this has two main implications. First, it is possible that more trees are cut for timbering but less minerals are obtained. Secondly, it may be related to marketing problems. The Zonal Mining Officer may not know the market for minerals, hence fail to collect enough revenue.

[33] The Chairman of Kigwasi Village, Umba, 1998

[34] The Regional Planning Officer, Tanga

environmental degradation. In the case of sand mining, the regional authorities have stopped issuing claim titles in areas that are prone to soil erosion. In the case of salt-mining along the coast, there is tight security provided by the Tanga Coastal Management Project (TCMP). Another measure is to encourage the use of solar pans in boiling and drying and to allow salt-mining only in areas without mangroves.

Impact on Water Resources

In the preceding sections, problems of water pollution have been reported from secondary data. In this section a few findings from the survey are presented to indicate the perception of respondents towards environmental impacts on water resources. Results from the structured questionnaire indicate that 49.3% of interviewed persons in the three areas do not believe that water is contaminated by mining activities compared with 36.6% who believe that Mwanza has the highest percentage of respondents (56.3%) who consider water to be contaminated, followed by Arusha (37.1%) and Tanga (20%). This result is explained by the differences in the type of minerals extracted in the three areas and the technology employed. Whereas gold-mining uses water and mercury at various stages in the production process, gemstones do not require water in the same way that gold-mining does.

Appendix 2: *Major Post-1950 Exploration Ventures*

Period	*Organization/ Company*	*Activity*	*Area*	*Minerals*
1947-1961	New Consolidated Goldfields	Exploration, mining	Various	Au, Sn, kaolin
1940-present	Williamson Diamonds Ltd.	Mining, exploration	Mwadui, Shinyanga	diamond
1959-1963	British Geological Survey	Exploration	Geita, Rwamagaza	Au
1965-1968	UNDP Special Fund	Exploration	Buck Reef Mine, greenstone bents	Au, Cu, Zn
1966	Hunting Surveys	Airborne geophysics	Musoma	Au,Cu,Zn
1967	Chinese Team	Feasibility	Songwe-Kiwira	Coal
1967	Chinese Team	Exploration	Tazara Railway	Various
1970	STAMICO/ GEOMIN (Romania)	Feasibility	Buck Reef	Au
1970-1974	MADIN/USSR	Exploration	Lupa/Mpanda	Au, base metals
1971	Williamson Diamonds Ltd.	U/g exploration	Buck Reef	Au
1973-1974	SIDA (Sweden)	Development	Buck Reef	Au
1974-1976	STAMICO/ Japan	Feasibility study	Lake Natron	Soda ash
1974-1976	STAMICO/ GEOMIN	Exploration	Coastal Belt	Beach sands
1974-1988	STAMICO/ China	Exploration, mining	Songwe-Kiwira coal field	Coal
1975-1978	STAMICO/ GTZ	Exploration, feasibility study	Mchuchuma Liganga	Coal Iron
1975-1979	Madini	Exploration	Mara	Asbestos, diamond
1976	STAMICO	Exploration, mining	Minjingu	Phosphate

Period	*Organization/Company*	*Activity*	*Area*	*Minerals*
1976-1979	UNDP	Exploration	West Lake region	Ni, Sn,etc.
1976-1980	MADINI/ STAMICO	Drilling	Bulyankhulu	Au, Cu
1977-1981	Sala International (Sweden)	Development	Buck Reef	Au
1977-1980	Geosurvey International	Airborne geophysics	Entire country	All minerals
1978-1979	STAMICO	Exploration	Bulyankhulu	Au, Cu
1978-1981	STAMICO/ AUSTROPLAN	Exploration, feasibility	Pugu	Kaolin
1979-1981	Madini	Exploration	Nzega, Geita	Au, Cu, Zn
1980-1981	Geosurvey International	Drilling, geophysics	Shinyanga	Diamonds
1980-1981	Geosurvey International	Airborne geophysics	Nzega, geita	Au, Cu, Zn
1980-1982	UNDP/ MADINI	Exploration	N.Tanzania	Ni, Cu, Co, U
1981-1982	TANEX (Williamson)	Exploration	Mwadui	Diamonds
1981-1986	STAMICo/ Finland	Drilling	Bulyankhulu	
1982	STAMICO/ Delft University	Inventory, exploration	Sekenke	Au,Cu
1982-1986	STAMICO/ Finland	Feasibility	Bulyankhulu	Au,Cu, Zn
1983-1987	UNDP	Exploration	Zanzui, geita Nyamulilima Canuck	Au,Ni, Cu,Zn
1984-1985	STAMICO/ Bulgaria	Exploration	Samena/geita	Au-Cu
1984-1986	STAMICO/ World Bank	Exploration	Songwe Kiwira	coal
1984-1991	Germany	Geology	Greenstones belts	Au,ni,PGM
1984-1989	Dar Tadine Al-Umma	Exploration	Greenstones belts	Au
1989-1992	Placer Dome	Exploration	Bulyankhulu	Au,Cu
1989-1996	Romanex International	Exploration	Kagera	Ni, Co
1991-1995	UNRFNRE	Exploration	Nyamulilima, Geita, Canuck (Nzega)	Au

Appendix 3: *Major Post-1950 Exploration Ventures*

Period	*Organization/ Company*	*Activity*	*Area*	*Minerals*
1947-1961	New Consolidated Goldfields	Exploration, Mining	Various	Au, Sn, kaolin
1991-present	Anglo American/ DeBeers (TANEX)	Exploration	Mwanza-Tabora region	diamonds
1991-1994	National Dev. Corporation	Exploration	Liganga	Fe
1991-present	TOFCO	Exploration	Various	gemstones
1992-present	New Era Development Ltd.	Exploration	Mabadala	Au
1992-present	Kilimanjaro Mines Ltd.	Exploration	Shinyanga	Au
1992-present	Dhahabu Expl. & Mining Ltd.	Exploration	Chunya	Au
1992-present	Sampo Resources (T) Ltd.	Exploration	Butiama & Ikungu	Au
1992-present	Tancan Mining Co. Ltd.	Exploration	Ngula, Luhala, Kibanda, Mwingilo, Mulehe, Mrekezi	Au, diamond
1992-present	Tuckilometersan Mines and Minerals Ltd.	Development	Merelani	tanzanite
1992-present	Madaba Mining Minerals Ltd.	Exploration	Kahama	Au
1992-present	Precious Minerals of Tanzania Ltd.	Exploration	Umba	ruby
1993-present	Afrika Mashariki Gold Mines	Development	Kahama, Tarime, Nyabirama	Au
1993-present	Twiga Mining Ltd.	Exploration	Nzega, Singida	Au
1993-present	Stilton Mining (T) Ltd.	Exploration	Igunga, Chunya, Singida	Au

Period	*Organization/ Company*	*Activity*	*Area*	*Minerals*
1993-present	Buibui Investments/ SAMAX	Exploration	Chunya	Au
1993-present	Mozan Ltd.	Exploration	Monduli	gemstones
1993-present	Shakaas Ltd.	Exploration	Chunya, Iramba	Au
1993-1994	Skeat Mining	Exploration	Matinje, Nzega	Au
1993-present	Global resources	Exploration	Iramba	diamonds
1994-present	Pangea Gold	Exploration	Bunda (Suguti), Kahama	Au
1994-present	Kisimamba Reef Mining Co.Ltd.	Exploration	Singida, Mbulu	Au
1994-present	Tabam Enterprises	Exploration	Singida	Au
1994-present	M.T.D. Mining Ltd.	Exploration	Ngasamo, Seregeti	Au
1994-present	Mwansiro Reef Ltd.	Exploration	Singida	Au
1994-present	Graphtan Ltd.	Production	Merelani	graphite, tanzanite
1994-present	Bakkar	Exploration	Singida, Songea	Au
1994-present	Kahama Mining Corp.	Development	Bulyankhulu	Au
1994-present	Ashanti G.m. (formerly Cluff)	Development	Geita	Au
1994-present	Eurotan Agricultural	Exploration	Tarime	Au
1994-present	Jumbo & Co. Ltd.	Exploration	Chunya	Au
1994-present	Parama and Co. Ltd	Exploration	Kasamwe,Igunga, Ikina	Au
1994-1996	Sutton Resources/BHP	Exploration	Kagera Region	Ni,Co

Period	*Organization/ Company*	*Activity*	*Area*	*Minerals*
1994-1996	Reunion Mining PLC	Exploration	Mabula, Lunguya	Diamonds
1994-present	Claremont Resources	Exploration	Lelatema	Diamonds
1994-present	SAMAX	Exploration	Ushirombo, Mabangu	Au
1995-present	RTZ	Exploration	Tabora	diamonds
1995-present	Ashanti/pangea JV	Exploration	South Bulyanhulu	Au
1995-present	Randgold and Expl/pangea JV	Exploration	Golden Ridge, South of Mwanza	Au
1995-present	Anglo-American/Pangea JV	Exploration	Kahama	Au
1995-present •	Mchuchuma (AREA)Coal	Development	Manda	Coal

Appendix 4: *Large Scale Mineral Rights Granted From 1990 to 1999*

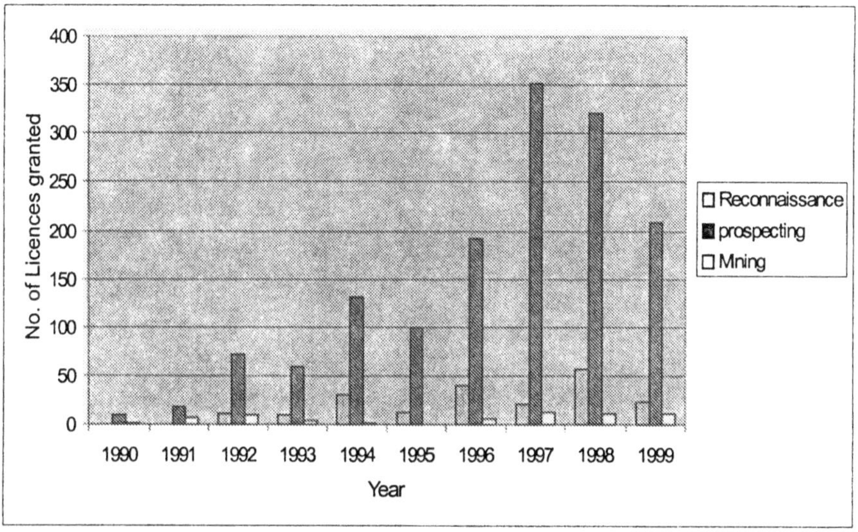

Source: Ministry of Energy and Minerals

Appendix 5: *Large Scale Mineral Rights Granted in 1997*

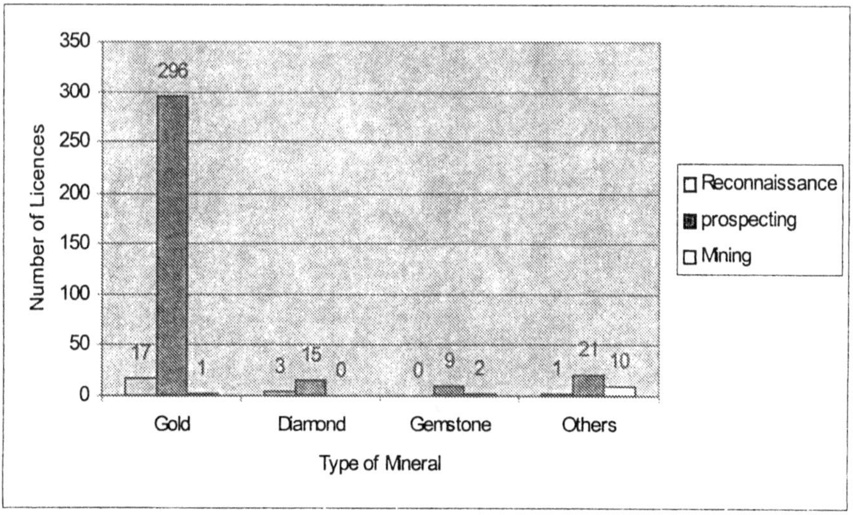

Appendix 6: *Status of Prospecting Licences Granted Since 1990 up to 30th June 1999*

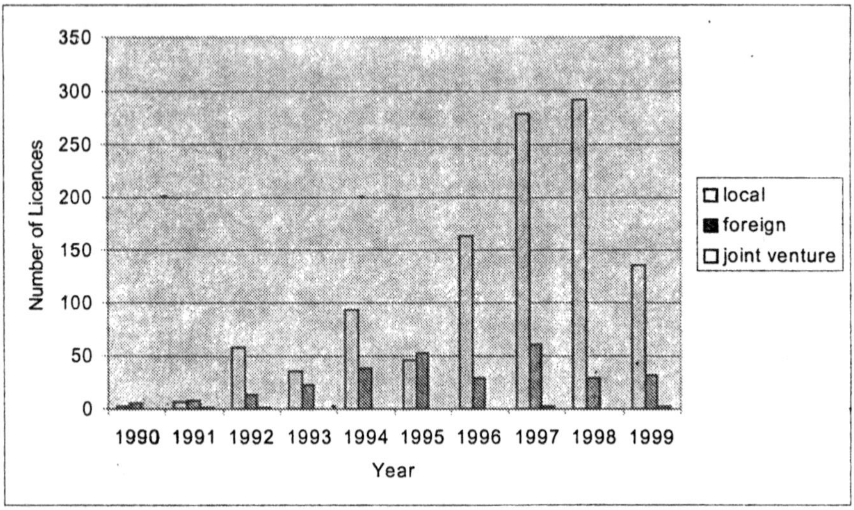

Appendix 7: *Status of Mining Licences Granted Since 1990 up to 30th June 1999*

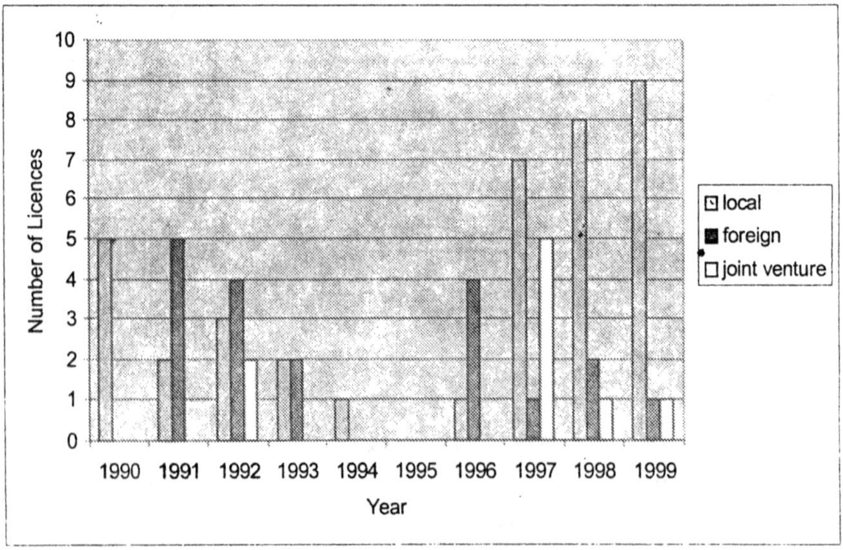

www.ingramcontent.com/pod-product-compliance
Lightning Source LLC
LaVergne TN
LVHW022037200926
848481LV00003B/711
* 9 7 8 9 9 7 6 6 0 3 9 0 3 *